T0074875

Cambridge Elements

Elements of Paleontology
edited by
Colin D. Sumrall
University of Tennessee

TESTING CHARACTER EVOLUTION MODELS IN PHYLOGENETIC PALEOBIOLOGY

A Case Study with Cambrian Echinoderms

April Wright
Southeastern Louisiana University

Peter J. Wagner
University of Nebraska, Lincoln

David F. Wright
National Museum of Natural History (Smithsonian Institution)

CAMBRIDGE
UNIVERSITY PRESS

University Printing House, Cambridge CB2 8BS, United Kingdom

One Liberty Plaza, 20th Floor, New York, NY 10006, USA

477 Williamstown Road, Port Melbourne, VIC 3207, Australia

314–321, 3rd Floor, Plot 3, Splendor Forum, Jasola District Centre, New Delhi – 110025, India

103 Penang Road, #05–06/07, Visioncrest Commercial, Singapore 238467

Cambridge University Press is part of the University of Cambridge.

It furthers the University's mission by disseminating knowledge in the pursuit of education, learning, and research at the highest international levels of excellence.

www.cambridge.org
Information on this title: www.cambridge.org/9781009048842
DOI: 10.1017/9781009049016

First published 2021

A catalogue record for this publication is available from the British Library.

ISBN 978-1-009-04884-2 Paperback
ISSN 2517-780X (online)
ISSN 2517-7796 (print)

Additional resources for this publication at www.cambridge.org/wrightwagnerwright

Testing Character Evolution Models in Phylogenetic Paleobiology

A Case Study with Cambrian Echinoderms

Elements of Paleontology

DOI: 10.1017/9781009049016
First published online: July 2021

April Wright
Southeastern Louisiana University
Peter J. Wagner
University of Nebraska, Lincoln
David F. Wright
National Museum of Natural History (Smithsonian Institution)

Author for correspondence: April Wright, april.wright@southeastern.edu

Abstract: Macroevolutionary inference has historically been treated as a two-step process, involving the inference of a tree, and then inference of a macroevolutionary model using that tree. Newer models blend the two steps. These methods make more complete use of fossils than the previous generation of Bayesian phylogenetic models. They also involve many more parameters than prior models, including parameters about which empiricists may have little intuition. In this Element, the authors set forth a framework for fitting complex, hierarchical models. The authors ultimately fit and use a joint tree and diversification model to estimate a dated phylogeny of the Cincta (Echinodermata), a morphologically distinct group of Cambrian echinoderms that lack the fivefold radial symmetry characteristic of extant members of the phylum.
Although the phylogeny of cinctans remains poorly supported in places, this Element shows how models of character change and diversification contribute to understanding patterns of phylogenetic relatedness and testing macroevolutionary hypotheses.

Keywords: cinctans, phylogeny, fossils, divergence time, macroevolution

ISBNs: 9781009048842 (PB), 9781009049016 (OC)
ISSNs: 2517-780X (online), 2517-7796 (print)

Contents

1 Introduction

Historically, drawing macroevolutionary inferences from phylogenetic trees has been a two-step process (Harvey and Pagel, 1991). First, a researcher would estimate a phylogenetic tree from a matrix of phylogenetic characters (typically morphological characters or molecular sequence characters). Then, they would use that tree (or a set of trees, such as a posterior sample) to fit a macroevolutionary model. Over the past decade, models that blend macro-evolutionary inference with phylogenetic inference have become increasingly common. For example, the fossilized birth–death process is used to estimate dated phylogenetic trees (Stadler, 2011; Heath et al., 2014). This process is usually implemented as a Bayesian hierarchical model, in which one model describes the process of character change for phylogenetic characters, one describes the distribution of evolutionary rates over the tree, and one describes the process of speciation, extinction, and sampling that led to the observed tree (see Warnock and Wright in this issue for a more complete discussion of this). In this Element, we describe an approach to fitting complex hierarchical models using a focal dataset of cinctan echinoderms.

We can divide macroevolutionary hypotheses into two non-mutually exclusive groups: those making predictions about origination and extinction dynamics, and those making predictions about rates and modes of trait evolution. The latter group includes hypotheses about shifts in rates of anatomical change and hypotheses about driven trends in which particular character states become more (or less) common over time. Hypotheses predicting such patterns come both from macroecological theory and from evolutionary-developmental theory, and thus span a range of basic issues including developmental, ecological, and physical constraints, and selection (Valentine et al., 1969; Valentine, 1980). Research programs dedicated to assessing shifts in rates and modes of anatomical evolution have been a staple of quantitative paleobiology since the early 1990s. Accordingly, anatomical character evolution models that describe the predictions of these different macroevolutionary hypotheses have important theoretical implications for these endeavors.

Phylogeneticists have long been interested in the same sorts of character evolution models, albeit for very different reasons. Hypotheses of phylogenetic relationships make exact predictions about character state evolution among taxa given observed data and models of character change (e.g., Kimura, 1980; Felsenstein, 1981; Hasegawa et al., 1985; Tavaré, 1986). The most common phylogenetic model for morphology makes the assumption of time-invariant models with no biases in the rate of character acquisition and loss (Lewis, 2001). The expectations of character evolution, of associations of characters with one another, and disparities between taxa are quite different when rates of acquisition and loss vary among characters and with time. This is particularly true when we include divergence times as part of phylogenetic hypotheses (Huelsenbeck et al., 2000; Sanderson, 2002; Drummond et al., 2006), but it is still true if we worry only about general cladistic relationships (i.e., which taxa are most closely related to each other; see Felsenstein (1981); Nylander et al. (2004); Wright et al. (2016)). In other words, many of the conceptual mice that paleobiologists seek to capture are the conceptual mousetraps that systematists seek to use to capture phylogenies.

Many readers' first instincts will be that this presents paleobiological phylogeneticists with a quandary: which comes first, the character evolution models or the phylogenetic inference? Part of the dilemma here stems from a historical view that we should treat phylogenetic analysis and macroevolutionary analysis as two separate endeavors (e.g., Harvey and Pagel, 1991). When we estimate a phylogenetic tree, we typically need to make simplifying assumptions about the evolution of our phylogenetic characters for tractability of the analysis. For our comparative methods, we are often using a smaller subset of the data to explore more complex models, possibly even seeking to falsify those same simplifying assumptions. Here, we advocate a very different approach that stems from hierarchical Bayesian phylogenetic approaches. That is, we should not view phylogenetic analysis and macroevolutionary analysis as two independent projects, but instead as two parts of the same endeavor of unraveling the evolutionary history of fossil taxa. These evolutionary histories include when clades and lineages diverged, the consistency of character change rates, biases in state acquisition, the process of diversification that led to the observed tree, and (of course) exactly how taxa were related to each other. Along the same lines, we have to accept and even embrace the fact that there will always be some degree of uncertainty in all of these things. These uncertainties are not a reason to abandon the endeavor as hopeless; on the contrary, it will mean that those conclusions that we can reach do not assume that specific historical details are true.

In this work, we will provide an example of the approach that we are advocating using a series of analyses of the Cincta, an extinct clade of "carpoid" echinoderms from the middle Cambrian. We will detail how paleobiologists can adapt different clock models and character state evolution models initially devised to accommodate uncertainties in molecular evolution to represent and model macroevolutionary hypotheses. In doing so, we will also outline protocol that paleontologists can replicate to conduct analogous analyses on other clades. We will emphasize how the combination of Markov Chain Monte Carlo analyses and stepping-stone tests allow us to marginalize specific details of character evolution models and phylogenetic relationships in order to generate the best joint summary of a clade's evolutionary history. Because there are innumerable possible models that one might consider, we will draw attention to existing methods with which paleontologists might already be familiar that should be useful for suggesting particular models as worthy of consideration. Finally, we will briefly outline other theoretical and methodological areas that remain for paleobiologists and systematists to resolve and unite in the future.

2 Taxonomic Background and Data

2.1 Cincta: An Enigmatic Clade of Cambrian Echinoderms

Echinoderms are a diverse phylum of marine animals represented today by more than 7,000 living species (Brusca and Brusca, 2003) distributed among five extant classes, including sea stars, brittle stars, echinoids, sea cucumbers, and crinoids. However, the spectacular diversity of extant echinoderms, measured by both species richness and anatomical variety, represents a paltry fraction of their prodigious evolutionary history recorded in the fossil record. Because echinoderms possess a mineralized endoskeleton made of high-magnesium calcite (calcium carbonate) and occur in virtually all habitats across the spectrum of marine depositional environments, the echinoderm fossil record is spectacularly complete and reveals approximately 30 clades distributed among 21 taxonomic classes spanning the entire Phanerozoic Eon (Sprinkle and Kier, 1987; Sumrall, 1997; Sumrall and Waters, 2012; Zamora and Rahman, 2014; Wright et al., 2017; Sheffield and Sumrall, 2019). Unlike familiar echinoderms inhabiting modern oceans, such as sea stars and sea urchins (echinoids), Cambrian lineages comprise an unfamiliar, taxonomically and morphologically diverse assemblage of predominately stem-group taxa exhibiting a diversity of body plans, life modes, and ecological traits unseen in extant lineages (Sprinkle, 1973; Zamora et al., 2013a; Zamora and Rahman, 2014).

Perhaps no group of early echinoderms has received greater attention and controversy than the carpoids (Rahman, 2009). Sometimes called homalozoans or calcichordates in the literature, carpoids comprise a heterogenous assemblage of extinct echinoderms including ctenocystoids, cinctans (Homostelea), solutes (Homoiostelea), and stylophorans. Although carpoids possess unique skeletal features that unambiguously identify their echinoderm affinities (David et al., 2000; Bottjer et al., 2006; Rahman, 2009; Zamora et al., 2020), they lack other traits considered synapomorphies of crown-group echinoderms. For example, all extant echinoderms exhibit pentaradial symmetry in adults and possess a water vascular system, unique to the phylum, used for locomotion, respiration, and excretion (Nichols, 1972). In contrast, 'carpoid' taxa exhibit bilateral to asymmetrical forms, and it's debated whether some possessed a water vascular system (Smith, 2005; Lefebvre et al., 2019). Although the phylogenetic position of carpoids has long been contentiously debated (see Rahman, 2009, and Rahman et al., 2009, and articles cited therein), only recently have computer-based phylogenetic analyses played a major role in evaluating alternative hypotheses (Sumrall, 1997; Smith and Zamora, 2013; Zamora and Rahman, 2014), and only one previous study tested phylogenetic hypotheses using stratigraphic data (Rahman et al., 2009). Crucially, the character matrices constructed for these analyses have greatly benefited from recent improvements to identifying homologous characters across morphologically disparate early echinoderm lineages, often arising from new fossil discoveries (e.g., Zamora et al., 2012; Smith and Zamora, 2013). Taxonomic controversy remains (David et al., 2000), though both recent computational phylogenetic analyses and stratigraphic congruence metrics support the hypothesis that carpoids comprise a paraphyletic assemblage of stem-group echinoderms (Rahman et al., 2009; Smith and Zamora, 2013; Zamora and Rahman, 2014). If this view is correct, then carpoids help document the radical transition in echinoderm evolution from an ancestral, bilaterian body plan to the pentaradial symmetry characteristic of crown-group forms that have dominated marine ecosystems since the close of the Cambrian. Regardless of their specific branching relationships in the echinoderm tree of life, it is nevertheless clear that understanding the distribution of character combinations and patterns of trait evolution in these enigmatic, pre-radial echinoderm lineages is critical to deciphering the early evolution of the phylum.

Cinctans are a significant group of non-radiate, carpoid echinoderms temporally restricted to the middle Cambrian (Miaolingian 509–497 Ma, with occurrences of cinctans from 506.6–297 Ma) and paleogeographically restricted to western Gondwana, Avalonia, and Siberia. Cinctans are generally small (i.e., 1 to 10 mm in length), flattened, symmetrical to irregularly shaped fossils resembling a tennis racquet or badly formed pancake, generally interpreted as an adaptation to an epibenthic, suspension-feeding lifestyle (Rahman and Zamora, 2009; Rahman et al., 2015). Like all echinoderms, cinctans have a complex, multielement, calcitic endoskeleton, which makes them particularly amenable for coding discrete, phylogenetic characters in fossil taxa (Smith and Zamora, 2009). The main body, called the theca, is surrounded by a series of rigid, stout, marginal plates (called the cinctus), which surrounds a central body of smaller, tessellated integument plates on both dorsal and ventral sides. The mouth is a circular opening located at the end of a narrow food groove (or pair of grooves) on the right anterior side of the theca. A posterior appendage,

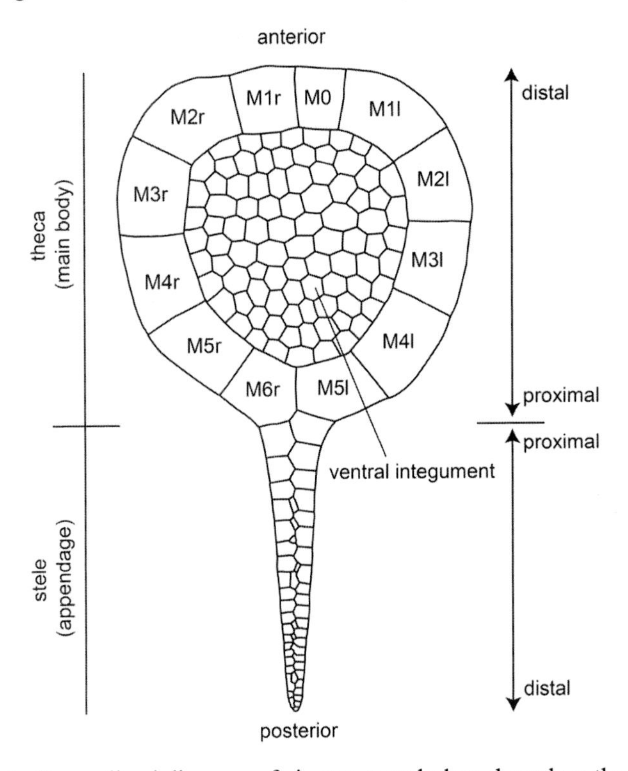

Figure 1 Generalized diagram of cinctan morphology based on the ventral side of *Trochocystites*. Marginal plates (i.e., comprising the cinctus) are labeled from anterior to posterior following Friedrich (1993); "l" and "r" refer to the left and right sides of the theca in dorsal view. See Smith and Zamora (2009) and Rahman (2016) for additional views of fossils and their anatomical reconstructions.

called the stele, forms a rigid structure extending from the cinctus, commonly subequal in length to the theca (Figure 1).

Despite their diminutive size, geological antiquity, and narrow paleogeographic and stratigraphic ranges, the significance of cinctans to understanding early echinoderm evolution, as well as their evolutionary implications surrounding ancestral character states in ancient deuterostomes (Smith and Swalla, 2009), has led to a substantial amount of interest to decipher their paleobiology. Recent advances in cinctan paleobiology include efforts to better understand patterns of taxonomic diversity (Zamora and Álvaro, 2010), ontogeny and development (Smith, 2005; Zamora et al., 2013b), life mode and feeding ecology (Rahman et al., 2009, 2015; Zamora and Rahman, 2015), convergence and adaptive evolution (Zamora and Smith, 2008), and phylogenetic relationships (Friedrich, 1993; Sdzuy, 1993; Smith and Zamora, 2009; Zamora et al., 2013b). In this Element, we combine morphological data with fossil age information to reevaluate phylogenetic relationships and evolutionary dynamics among cinctan lineages using hierarchical Bayesian phylogenetic models, and provide a phylogenetic template for future systematic and macroevolutionary studies.

2.2 Character Data

We use the character data initially published by Smith and Zamora (2009) and subsequently augmented by Zamora et al. (2013b). The analyzed matrix includes 22 cinctan species plus one outgroup taxon (*Ctenocystis*, represented by *Ctenocystis utahensis*). An additional four cinctan species are excluded due to inadequate material for coding. We refer the readers to the papers cited above for additional information concerning the character data.

Both disparity analyses of these data conducted and arguments pertinent to early echinoderm evolution in the literature (e.g., Smith et al., 2013) suggest that cinctans might exhibit "Early Burst"-type dynamics (Figure 2A), in which a broader range of anatomies appears early in clade history than expected if rates of change were reasonably consistent over time. Standing disparity versus log-standing richness patterns deviate strongly from expectations given constant rates of change (Jablonski, 2020; Wright, 2017a), but this reflects in part the clade decreasing in richness through its later history. The same relationship with cumulative disparity (i.e., disparity among all species known by some date) shows a weaker trend toward higher disparity than expected during the first half of clade evolution (Figure 2C). This in turn suggests that rates of change might have been higher early in clade history (Foote, 1996b).

2.3 Chronostratigraphic Data

Our chronostratigraphic data come from 221 occurrences of Cambrian echinoderm species from 143 localities last downloaded from the Paleobiology Database (PBDB) on January 1, 2020 (Wagner, 2021).

The locality and occurrence data came from 81 references with the seven biggest sources including Nardin et al. (2017), Zamora (2009), Chlupac et al. (1998), Sprinkle and Collins (2006), Termier and Termier (1973), and Sprinkle (1973). We ourselves entered 108 of those occurrences and 72 of those localities, and updated 51 of the remaining localities for the purpose of this Element. After accounting for synonymies and variant spellings, the localities represent 55 different rock units (i.e., formations and formations+members). The accepted names of the species occurrences reflect 361 taxonomic opinions, 155 of which we entered for the purposes of this Element.

The Paleobiology Database returns ages based only on the entered interval, which usually is a stage. Here, nearly every cinctan-bearing locality is assigned

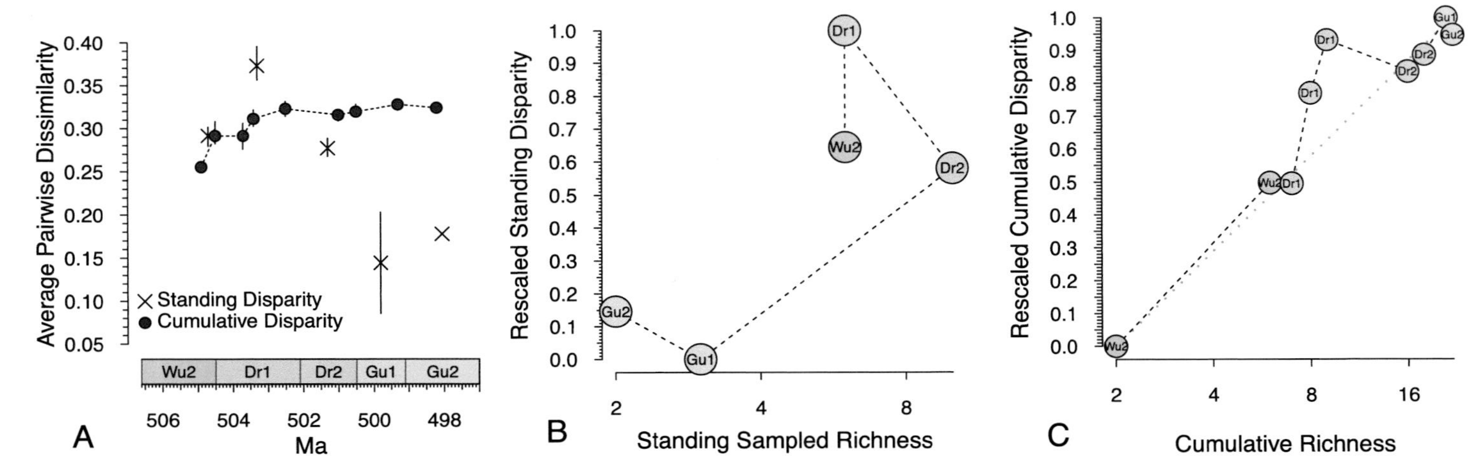

Figure 2 Disparity patterns for cinctans. (A) Disparity over time based on average pairwise dissimilarities among taxa (Foote, 1992). Vertical bars represent 90-percentile error bars from bootstrapping pairwise comparisons (Foote, 1993). X gives "traditional" standing disparity, which reflects only species extant during a stage slice (see, e.g., Hughes et al., 2013). Note also that a minimum of two species must be present. Circles give the cumulative disparity among all members of the clade sampled in rocks that age or older, regardless of whether they are still extant, and thus depict the total range of anatomical types derived within the clade (Wagner and Estabrook, 2015). (B) Rescaled standing disparity versus log-taxonomic richness. The dashed line here and in (C) gives the expected change given continuous rates of morphological innovation. Because cinctans begin with relatively high richness and then decline over time, the curve begins near the middle of the plot rather than near the bottom as is typical (Jablonski, 2020). Note that initial disparity increases markedly despite no increase in standing richness. (C) Rescaled cumulative disparity given cumulative richness. The scale here is finer as it reflects disparity evolved through points in time rather than extant during stage slices. The rapid rise in disparity from the Wuliuan to the early Drumian in (A) and (B) now reflects the appearance of very anatomically disparate species in the early Drumian while earlier species introduce disparity comparable to that introduced by late Drumian and later species.

to the middle Cambrian and thus receives a possible age of 513–501 Ma. However, PBDB provides information allowing much more exact ages. For example, PBDB collection 67775 is one of four including *Trochocystites bohemicus*. This collection is assigned to the Middle Cambrian and thus is dated by the PBDB as 513–501 Ma. However, this collection represents the Skryje Shale, which is known to span four trilobite zones that restrict the age to 505.2–500.7 Ma (based on correlations among trilobite zones by Geyer (2019) to trilobite zone ages in Gradstein et al. (2012)). Thus, if we had no further information, then those would be the oldest and youngest possible ages for this collection. However, PBDB collection 67775 also is assigned to the *Eccaparadoxides pusillus* trilobite zone, which further restricts the age to 505.2–504.5 Ma. We use a database of Paleozoic rock units and faunal zones compiled by one of us (PJW) as a thesaurus to provide more exact earliest and latest possible ages for each cinctan-bearing collection. In addition to refining dates, the database also effectively updates the chronostratigraphic unit to which a locality is assigned if current ideas about the age of a trilobite zone have changed since the paper(s) providing the original data. The typical locality now can be restricted to a 0.7 million year window. Prominent sources for the information relevant to our Element and for interregional correlations of rock units and trilobite zones include Alvaro et al. (2001); Liñán et al. (2004); Geyer and Landing (2006); Geyer and Shergold (2000); Alvaro et al. (2007); and Geyer (2019). The overall timescale is that of Gradstein et al. (2012). (Note that the more recent timescale of Gradstein et al. (2020), which was published after we conducted these analyses, provides nearly identical dates for the relevant Middle Cambrian trilobite zones and thus should not generate markedly different results.)

We use the refined dates to put lower and upper bounds on the possible first-appearance (FA) dates of the cinctan species (Table 1). For species known from only single intervals or trilobite zones, the lower and upper bounds for both first and last appearances are necessarily identical. This is not the case for species spanning multiple intervals. For example, *Gyrocystis platessa* occurs in rocks as old as the *Badulesia granieri* trilobite zone (504.9–504.5 Ma given Geyer (2019) and Gradstein et al.'s timescale) and also occurs in rocks as young as the *Solenopleuropsis marginata* trilobite zone (501.0–499.3 Ma). Here, the latest first possible appearance is 504.5 Ma. We choose the widest possible uncertainty. For example, *Gyrocystis erecta* occurs in rocks belonging to the *Solenopleuropsis* zone (503.1–501.0 Ma) but also in rocks dated more specifically to the *Solenopleuropsis thorali* subzone (501.6–501.0 Ma). Because the former set of occurrences might be as old as 503.1 Ma (given existing information), we set the possible lower and upper bounds on the FA for *G. erecta* at FA_{LB}=503.1 and FA_{UB}=501.0 Ma.

Table 1 Chronostratigraphic information for analyzed taxa based on occurrences in the Paleobiology Database. FA and LA denote first and last appearance dates, with LB and UB giving the oldest and youngest possible FA and LA given the finest chronostratigraphic resolution possible (e.g., a trilobite zone or local chronostratigraphic unit). "N_S" gives number of sites (= collections or localities) that a species occupies. "N_R" gives number of rock units (formations or members) that a species occupies. Dates for *Ctenocystis* represent the entire genus; however, the coded species (*C. utahensis*) is also the oldest known *Ctenocystis*.

Taxon	FA_{LB}	FA_{UB}	LA_{LB}	LA_{UB}	N_S	N_R
Ctenocystis	506.6	506.5	501.0	500.5	4	3
Gyrocystis platessa	504.9	504.5	501.0	499.3	13	4
Gyrocystis testudiformis	503.1	502.5	503.1	502.5	4	1
Gyrocystis cruzae	503.1	501.0	503.1	501.0	1	1
Gyrocystis badulesiensis	503.1	501.0	503.1	501.0	2	1
Gyrocystis erecta	503.1	501.0	501.6	501.0	2	1
Progyrocystis disjuncta	503.1	501.0	503.1	501.0	1	1
Protocinctus mansillaensis	506.6	505.4	506.6	505.4	1	1
Elliptocinctus barrandei	501.6	501.0	501.6	499.3	6	3
Elliptocinctus vizcainoi	504.5	503.4	504.5	503.4	1	1
Sucocystis theronensis	501.6	501.0	501.6	501.0	2	2
Sucocystis bretoni	501.0	500.5	501.0	500.5	1	1
Lignanicystis barriosensis	501.6	501.0	501.6	501.0	3	1
Undatacinctus undata	501.0	499.3	501.0	499.3	1	1
Sucocystis acrofera	499.3	498.2	499.3	498.2	2	1
Undatacinctus quadricornuta	501.0	499.3	501.0	499.3	1	1
Undatacinctus melendezi	501.0	499.3	498.2	497.0	9	2
Asturicystis jaekeli	504.9	504.5	504.9	504.5	1	1
Sotocinctus ubaghsi	505.4	504.9	505.4	504.5	2	2
Trochocystites bohemicus	505.2	504.5	503.0	502.2	4	3
Trochocystoides parvus	504.5	503.7	504.5	503.7	1	1
Ludwigicinctus truncatus	501.6	500.5	501.6	500.5	1	1
Graciacystis ambigua	504.9	504.5	503.7	503.1	3	1

3 Methods

3.1 Estimating Starting Values for Sampling and Diversication Rates

We use the Paleobiology Database occurrences described above to derive initial estimates of origination, extinction, and sampling for Cambrian echinoderms. Although this Element focuses on just cinctans, other echinoderms represent a taphonomic control for initial estimates of sampling: rocks from environments permitting other identifiable fossils of other Cambrian echinoderms are those in which there is some probability > 0 that we would be able to sample cinctans if they had lived in those environments (Bottjer and Jablonski, 1988); in contrast, fossiliferous localities lacking identifiable echinoderms might represent environments in which echinoderms (cinctan or otherwise) lived, but that no longer represent sampling opportunities. Other echinoderms also provide a much larger sample size for initial estimates of origination and extinction rates than do cinctans. Although diversification rates likely varied among and within echinoderm clades as well as over time, paleontological data long have suggested that different major clades are typified by general diversification rates (Sepkoski, 1981). The larger sample sizes afforded by all echinoderms reduce the chance that our *initial* estimates will be wildly inaccurate as an artifact of sample size. We use a modified version of the Three-Timer method (Alroy, 2015) that uses lognormal distributions for sampling rates per stage slice rather than a single value (Wagner and Marcot, 2013). Note that we use both the diversification and sampling rates to seed the prior distribution with feasible starting values from which to generate new proposals for diversification and sampling parameters in MCMC generations and not as fixed values. Our rationale for adding this extra step is simply that, like all other search algorithms, MCMC analyses beginning with "realistic" parameters should converge on "correct" parameters faster than those analyses beginning with unrealistic parameters; and as diversification and sampling are among the parameters being varied from one iteration to the next in each analysis, this should make it easier for the analyses to achieve convergence for all of the parameters. We did estimate two parameters that remain static in MCMC analyses directly from PBDB data. One is the probability of taxon sampling for the youngest species. This parameter is separate from the general sampling parameter because, unlike this analysis, many analyses include extant taxa that reflect a fundamentally different sampling regime. The other is the earliest possible divergence date for the clade. We estimate this using the cal-3 metric (Bapst, 2013), with the lower bounds set at $p=0.003$ (i.e., $1 - 0.05^4$).

3.2 Models

The fossilized birth–death is a hierarchical model, meaning that different model subcomponents explain the evolution of the phylogenetic characters (the morphological evolution model), the distribution of evolutionary rates across the tree (the clock model), and the model that describes the speciation (λ), extinction (μ), and fossil sampling intensity (ψ) leading to the tree (the tree model).

Below, we describe a hierarchical approach to model-fitting, in which we fit a model to each subcomponent. The model subcomponents are then assembled into a total fossilized birth–death process.

For each model subcomponent, we first ran an MCMC in RevBayes to assess how long it takes for the analysis to reach convergence. Then, using this value, we ran 20 stepping-stone replicates to calculate a marginal likelihood for the data. Stepping-stone model-fitting samples iteratively in the space between the prior and the posterior. The aim in doing this is to estimate the probability of the data summed over all possible values for parameters (Xie et al., 2011). This enables the calculation of an unbiased marginal likelihood, in contrast to MCMC, which will be biased toward regions of treespace that contain good solutions.

The result of each stepping-stone analysis is a marginal likelihood. Because phylogenetic likelihoods tend to be quite small, they are typically reported as log-transformed values. This means that for model comparisons, we used the log Bayes Factor (Kass and Raftery, 1995), which is represented by the character K and given via the formula:

$$K = ln[BF(M0, M1)] = ln[P(X_j M0)] - ln[P(X_j M1)].$$

In the above equation, the Bayes Factor for model comparison between model 0 and model 1 is equal to the probability of the data (X) multiplied by model 0 minus model 1. The final Bayes Factor is calculated by exponentiating K:

$$BF(M0, M1) = e^K.$$

The final Bayes Factor is a single value for which a value greater than one constitutes support for model 1 and a value less than negative one is support for model 0.

Within each model subcomponent, Bayes Factors were used to compare different candidate models. The winning candidate model for each subcomponent was then used to estimate the subsequent FBD trees.

3.2.1 Morphological Evolution Models

We first fit a morphological character model, as no tree can be estimated without one. We compared three models for morphological character evolution. All three were based on the basic Mk model (Lewis, 2001). In this model, it is assumed that any character has an equal probability change and reversal between any two states. The data matrix was partitioned according to the number of character states, so that size of the transition matrix of the model was correctly specified. In the first model we did not allow rate heterogeneity.

In effect, this means we assume all characters to have the same rate of evolution. In the second, we used Gamma-distributed rate heterogeneity to allow different characters in the matrix to have different evolutionary rates. Smith and Zamora (2009) explicitly identify characters related to the "food groove" anatomy of cinctans. A common concern is that characters directly involved with basic organismal ecology and function such as those involved in feeding might evolve under different rules than do other characters (Foote, 1994, 1996a; Wagner, 1995; Sánchez-Villagra and Williams, 1998; Ciampaglio, 2002; Hopkins and Smith, 2015; Wright, 2017a). Obviously, functional morphology of completely extinct groups such as cinctans resides in the realm of hypothesized rather than observed, even if those hypotheses can be corroborated (e.g., Rahman et al., 2020). This situation is akin to knowing that cinctans achieve high disparity rapidly. In that case, we have evidence for rate heterogeneity over time and/or among branches, and not allowing for this should make it more difficult to infer phylogeny accurately. Here, we have reason to suspect that two sets of characters are evolving at different rates, and not allowing for this also should make it more difficult to infer phylogeny accurately. Therefore, we executed partitioned analysis in which feeding and nonfeeding characters both have their own Gamma-distributed rate variation parameter, which should generate more probable overall hypotheses than analyses in which both partitions share the same Gamma-distributed rate variation parameter. Just as rejecting strict clock models in favor of early-burst or relaxed clock models informs us about more than phylogenetic relationships, rejecting a single partition in favor of "feeding versus nonfeeding" corroborates Smith and Zamora's original interpretation of the characters as well as refining our phylogenetic inferences. Conversely, failing to find a meaningful difference suggests either that the interpretations are incorrect or that cinctans are a group in which feeding and nonfeeding characters evolve at similar rates. Our "feeding versus nonfeeding" partitioned analysis doubles the total number of parameters by applying each to both character sets independently.

Our MCMC analyses reached convergence after about 80,000 generations, as checked in the software Tracer (Rambaut et al., 2018). Stepping-stones should generally be run to approximate convergence per stone. Therefore, each stepping-stone was run for 100,000 generations to account for any late-converging stones.

3.2.2 Clock Models

A phylogeny cannot be estimated without a model of character evolution. Hence, the morphology model was fit first. Next, we fit a clock model.

Although "clock" might conjure images of a constant rates model, that is only one type of clock (i.e., a "Strict Clock"): clock models include a range of models that either directly predict or at least constrain plausible rates of change. Without a morphology model, no tree can be estimated. Without an FBD model, age information cannot be included. Therefore, in order to fit a clock model, we used our best-fit morphology model and a simple, time-homogeneous FBD model to compete different clock models against one another.

The four candidate clock models were as follows.

- A strict clock: In this model, the rate of evolution along each branch is assumed to be equal. The rate of evolutionary change is sampled from an exponential distribution. Note that the strict clock model is the most simple clock model and also (an) equivalent to null models used in paleontological rate studies. It assumes that all branches follow a single, constant rate of morphological evolution. Although simplistic, some studies have found a surprising degree of concordance with fossil data fitting a strict morphological clock even when models incorporating rate variation provide a better statistical fit (Drummond and Stadler, 2016; Wright, 2017b). The strict clock model has one advantage in its simplicity: it adds only one parameter to the analyses, whereas relaxed clock models require many additional parameters. This clock can be thought of as a null model of evolutionary rate variation.
- An uncorrelated lognormal clock: This clock treats each branch as an independent draw from a distribution (Drummond et al., 2006; Drummond and Rambaut, 2007). In this case, we used a lognormal distribution, which says most evolutionary rates are likely to be low, but with allowances for some branches to have very high rates. It should be noted that because each branch is a separate draw, the rate of an ancestral branch's evolution may be very different than its descendants – either greater or lesser. In terms of macroevolutionary theory, an uncorrelated clock model is consistent with there being no shifts in intrinsic constraints on rates of change within a clade, but where there is considerable heterogeneity in the effects of ecology on rates of change among different lineages.
- Autocorrelated clock: These clocks assume that the rate of evolution on a descendent branch is drawn from a distribution centered on the rate of evolution of that branch's ancestor (Aris-Brosou and Yang, 2002). Here, the rate is heritable, but constantly changing in a manner analogous to a morphometric character under continuous change. Theoretically, we might expect this if rates are affected by some other variable that is continuously changing (e.g., climate variables or biological variables such as metabolism or size;

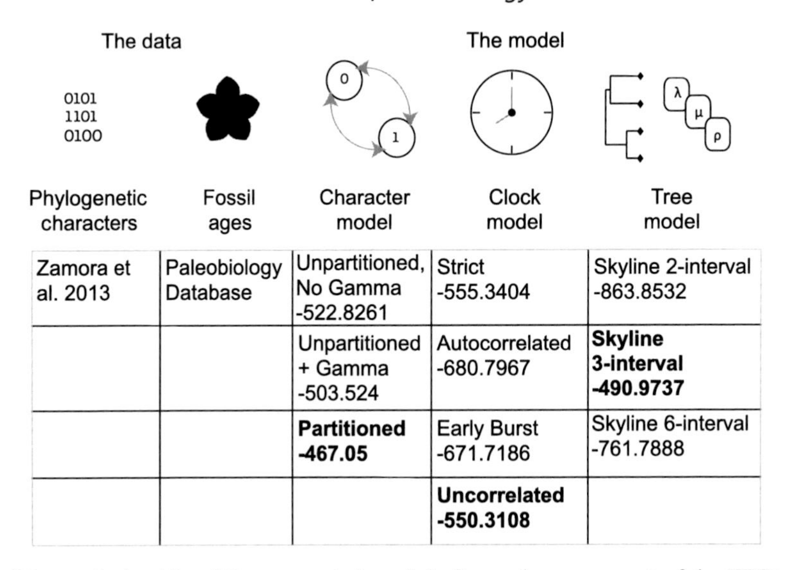

Phylogenetic characters	Fossil ages	Character model	Clock model	Tree model
Zamora et al. 2013	Paleobiology Database	Unpartitioned, No Gamma -522.8261	Strict -555.3404	Skyline 2-interval -863.8532
		Unpartitioned + Gamma -503.524	Autocorrelated -680.7967	**Skyline 3-interval -490.9737**
		Partitioned -467.05	Early Burst -671.7186	Skyline 6-interval -761.7888
			Uncorrelated -550.3108	

Figure 3 A table of the competed models for each component of the FBD process, and constituent parameters speciation (λ), extinction (μ), and fossil sampling intensity (ψ). Underneath each model component are the models competed for that component. The model indicated in bold text is the one that fit the data best, per Bayes Factor model selection. Warnock and Wright 2020. Cambridge Elements of Paleontology.

see further discussion in Bromham et al. (1996); Gaut et al. (1992); Thomas et al. (2006); Bromham et al. (2015)). Thus, this will favor smaller rate shifts than those seen in an uncorrelated clock. The amount of change expected between ancestor and descendant was modeled with a lognormal distribution. This assumes most descendants will have a similar evolutionary rate to their ancestors, but allows for some to have a larger disparity.

- Early Burst: This clock model treats rates of character change as exponentially decaying over time. This assumes that rates of evolution are fastest near the base of the tree and decline into the present. As illustrated above, disparity patterns within the clade also suggest this (Figure 1). Prior work has sought the question of detecting radiation in a phylogenetic context (Liow et al., 2010). This model builds on this idea while explicitly including fossils (Quental and Marshall, 2009, 2010).

Each of these models has a different number of parameters and took a different amount of time to converge. Therefore, for each model, we first ran an exploratory MCMC to see how long convergence takes. Then we used the convergence value to choose the number of iterations per stepping-stone. A table of competed models can be seen in Figure 3.

3.2.3 Tree Models

In all of our comparisons of tree models, we used variants of the fossilized birth–death model. We competed several models, reflecting different scenarios of diversification and sampling in the group. The simplest model treats these rates as constant over time. Of course, innumerable paleobiological studies indicate that origination, extinction, and sampling all vary over time within clades. Shifts in these rates mean that the prior probability of a branch spanning a given amount of time is not constant throughout clade history (Wagner, 2019). Skyline models (e.g., Stadler et al., 2013) treat this as a possibility within FBD analyses by allowing all three rates to vary in different time intervals. We contrasted several skyline models. These models assume that the parameters of the FBD analysis can vary between discrete time bins. The cinctan fossil record spans three geological stages of the middle Cambrian: the Wuliuan, Drumian, and Guzhangian. Most of the known species appear in the late Wuliuan and Drumian, with a marked decrease in the Guzhangian (Table 1). This suggests temporal variation in origination and/or extinction rates. Therefore, we allowed all analytical parameters to vary between all three geological stages. It should be noted that for all skyline models, there is an additional interval of time from the origin to the first interval with its own possible origination, extinction, and sampling rates.

- Time-homogeneous: The first FBD model is a time-homogeneous model in which it is assumed that one rate of speciation, extinction, fossil sampling, and sampling at the last occurrence time apply to the whole tree. Note that in Figure 3 every analysis in the "clock model" column used the time-homogeneous FBD because we need an FBD model in order to incorporate fossils.
- Two intervals: We tested a model in which the Drumian stage is split into two stages, for a total of two skyline categories (Wuliuan & Drumian 1, Drumian 2 & Guzhangian).
- Three intervals: We tested a model in which each stage is given its own set of FBD parameters, for a total of three skyline categories.
- Six intervals: In this model, we allowed each stage slice to have its own rates.

Although most prior FBD analyses treat origination and extinction as independent variables, paleobiological studies show that the two are closely correlated (e.g., Marshall, 2017). Over long periods of time, the relationship is nearly linear within large clades (e.g., Figure 4A, illustrating prominent Cambrian to Silurian clades). There is more variation with clades over short periods of time

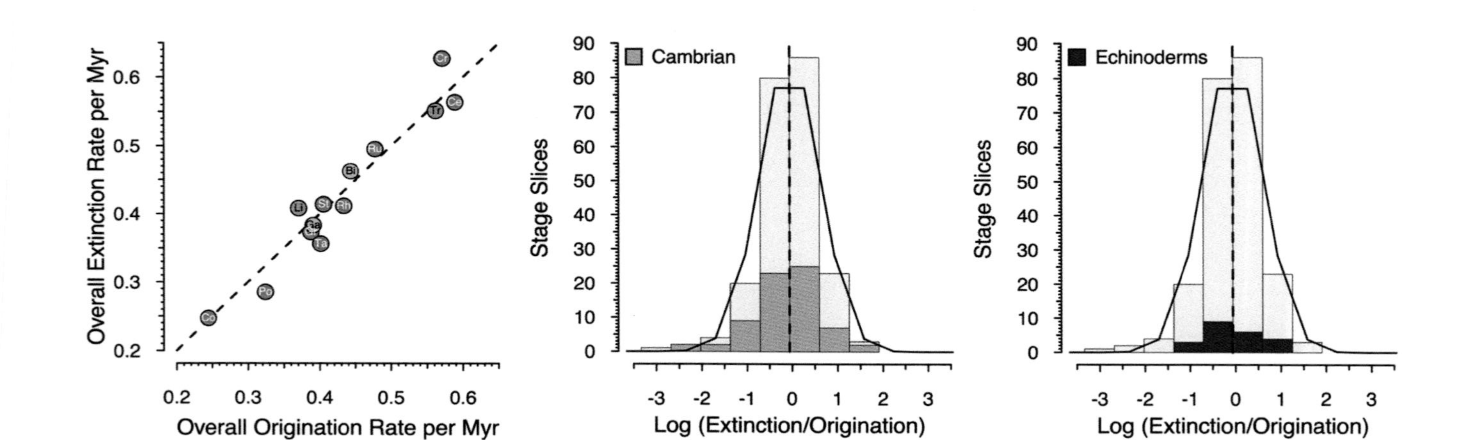

Figure 4 Relationships between origination and extinction rates given birth–death sampling analyses given Cambrian to Silurian data from the Paleobiology Database. (A) Most likely origination and extinction rates for species over entire histories of prominent higher taxa: trilobites (Tr), linguliform brachiopods (Li), conodonts (Co), poriferans (Po), tabulate corals (Ta), rugosan corals (Ru), rhynchonellate brachiopods (Rh), strophomenate brachiopods (St), cephalopods (Ce), crinoids (Cr), gastropods (Ga), and bivalves (Bi). Olive green denotes members of Sepkoski's Cambrian fauna (Sepkoski, 1981), dark green denotes members of Sepkoski's Paleozoic fauna, and light blue denotes members of Sepkoski's Modern fauna. (B & C) Distribution of logged turnover rates for individual clades (including but not restricted to the 12 groups in Figure 3(A) and individual stage slices (see Bergström et al. (2009), Cramer et al. (2011), and Rasmussen et al. (2019) for stage slice definitions.) The best-fit lognormal distribution also is illustrated. (B) Turnover rates for Cambrian time slices separated. (C) Turnover rates within echinoderm classes separated.

such as the stage slices that we use for these analyses (Figure 4B–C). However, there is still a distinct lognormal relationship between origination and extinction. Moreover, the lognormal relationship for only Cambrian stage slices or for echinoderms by stage slice fits the same overall lognormal relationship well. Thus, our MCMC searches vary origination as an independent variable and vary turnover (extinction/origination) as a variable dependent on origination. For the time-homogeneous model, this reflects the linear relationship shown in Figure 4A (where turnover varies from 0.90 to 1.05 times origination). For skyline models, turnover follows the lognormal relationships illustrated in Figures 4B–C.

Finally, for all competed models, the best-fit character change model and clock model were used as the other model subcomponents.

4 Results

4.1 Model Fitting

Model selection supported the choice of substitution model with feeding and nonfeeding characters (posterior probability: −467.053) modeled separately (log Bayes Factor: 2.938). It should be noted that Bayes Factor support values may appear small, but still reflect significance per the scale in Kass and Raftery (1995). An uncorrelated lognormal clock (posterior probability: −550.311) was favored over an autocorrelated clock (posterior probability: −680.797), a strict clock (posterior probability: −555.3404), or "early burst" dynamics (posterior probability: −671.719) with a log Bayes Factor of 4.768 (substantial evidence). Finally, the three-time interval model (posterior probability: −490.9737) was supported over the two-interval (−863.8532) and six-interval (−761.7888) models (Bayes Factor 5.809, substantial evidence). A schematic of the model competed with the best-fit models highlighted can be seen in Figure 3. Parameters of the best-fit FBD model can be seen in Table 2.

4.2 Cinctan Phylogeny

The phylogeny estimated can be seen in Figure 5. As expected, *Ctenocystis* is sister to the cinctans. *Protocinctus*, which has been recovered in some recent studies as nested deep within the cinctan clade (Smith and Zamora, 2009), is recovered here as sister to the rest of the clade. The genus *Gyrocystis* is monophyletic, with several species placed as sampled ancestors within the clade. *Progyrocystis* appears in a clade with *Asturicystis* and *Graciacystis*. This clade is sister to the *Gyrocystis*, albeit with low posterior support. *Trochocystoides* and *Trochocystites* are nested deeper in the tree than in prior analyses

Table 2 Diversification parameters of the FBD model. Rates presented as the median of the 95 percent HPD of the Bayesian posterior sample for the best-fit model, the model in which each geological stage has its own speciation (λ), extinction (μ), and fossil sampling intensity (ψ) in parameters. Turnover = μ/λ; Diversification=$\lambda - \mu$. A table with 95 percent HPDs for each value can be seen in Table A1.

Stage	ψ	Diversification	λ	μ	Turnover
Guzhangian	0.188	−0.193	0.493	0.687	2.148
Drumian	0.260	0.317	1.28	0.964	0.714
Wuliuan	0.224	−2.657	0.916	3.574	4.936

and are more closely related to species within the Sucocystidae. Similar to prior studies (Smith and Zamora, 2009; Zamora et al., 2013b), we also recover a clade comprising the genera *Sucocystis*, *Lignanicystis*, and *Ellipticinctus*, though in this analysis *Sucocystic acrofora* groups with the *Undatacinctus-Ludwigicinctus* subclade of Sucocystidae. The HPD on the age of the origin of Succocystidae is between 505.365 and 507.724 Ma.

5 Discussion

5.1 Model-Fitting for Complex Hierarchical Models

When the fossilized birth–death model was first implemented for divergence time estimation, one of the noted benefits was avoiding incoherent fossil calibration points on nodes (Heath et al., 2014). "Incoherent" here has multiple meanings: first, that fossils are not data under node calibration methods. In a node calibration framework, fossils constrain the possible ages a split can have. The fossil age ranges are not data under this framework (Gavryushkina et al., 2017). The researcher parameterizes what they believe to be the waiting time between the divergence and the fossil subtending it. This waiting time is capturing two different quantities – the uncertainty around the age of the fossil and how long since the divergence the fossil took to arise. In practice, choice of prior is often subjective and not based on any one criterion or method, though methods for doing this do exist (Nowak et al., 2013).

The second way in which this practice can result in incoherence is through the collision of priors on different nodes. Depending on the shape of the prior chosen, the upper age bound of an ancestor split may conflict with the lower age bound of its descendant splits. For example, if a researcher has little intuition for when a fossil arose in relation to the split that it subtends, they may place a uniform distribution specifying the longest and shortest waiting time between

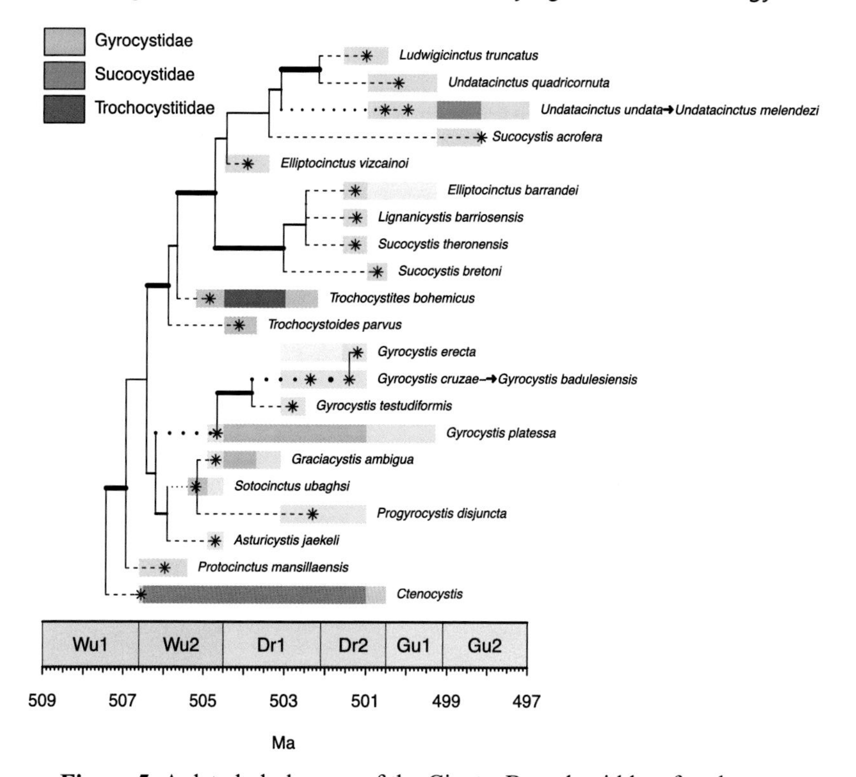

Figure 5 A dated phylogeny of the Cincta. Branch widths of nodes are proportional to the posterior probability of the branch, with wider branches reflecting higher posterior probability. Branch durations preceding sampled species are dashed lines; dashed nodes denote cases where we reconstruct a sampled species as ancestral to the others. We reconstruct instances where the implied ancestor was still extant when the daughter lineage appears as evidence for budding cladogenesis (see, e.g., Eldredge, 1971). Asterisks denote most probable ages of first appearances. For taxon ranges, solid colors reflect ages for which species have definitely older and definitely younger finds; transparent bars represent range of possible first and last appearances. Extra pale bars represent cases where some of the candidates for oldest/youngest occurrence might be that age. For example, some occurrences of *Gyrocystis erecta* might be 503.1–501.0 Ma whereas others might be 501.6–501.0 Ma.

the split and the fossil subtending it may be. Imagine this split and fossil are the descendants of an earlier node, which also has a fossil associated with it. Perhaps the researcher has an intuition that this older fossil is likely close to its ancestor node. And so the researcher places a lognormal prior on the fossil waiting time, saying the fossil is likely close to the node, but allowing for it to possibly be much older. If incorrectly parameterized, the upper bound of the lognormal could overlap the lower bound of the uniform, implying in those cases that the descendant split could occur before the ancestor split.

This is obviously undesirable. The fossilized birth–death model does not use node calibrations, instead parameterizing the uncertainty of the age associated with each tip. This is done by placing a uniform prior on each fossil tip that begins with the first occurrence of the fossil and ends with the last occurrence. For some taxa, this will mean a fairly wide uncertainty per tip. For example, in the terrestrial realm, fossil insect occurrences are often dated based on the type of amber they were found in (LaPolla et al., 2013). Some ambers can be precisely dated, as the trees that generate the amber have a narrow range. For others, the range of dates might be quite broad, as the amber type could be made from multiple trees, or in tree types with long geological persistences (Poinar and Mastalerz, 2000). In fossils that have been individually dated, this uncertainty may correspond to the uncertainty on the radiometric dating. Similarly, fossil age uncertainty is ubiquitous in the marine fossil record, even for well-sampled fossil taxa. For example, few marine fossils are sandwiched between rock units available for fine-scale radiometric dating. Instead, these layers must be correlated to other units using chronostratigraphic methods, which always involve an envelope of uncertainty. Sometimes, the oldest fossil belonging to a particular species may occur just above an unconformity (e.g., a sequence boundary), or stratigraphic correlations of fossil-bearing formations may otherwise be highly uncertain and contentious. Moreover, some fossils, particularly those from historical collections, may have been collected from a locality with low-precision stratigraphic data (i.e., "Silurian"), with no further information available to narrow its stratigraphic or temporal precision. Regardless of the manner in which the uncertainty is derived, the meaning is clear and singular: the uncertainty on a fossil tip represents the minimum and maximum plausible age of the fossil. This is a far clearer quantity to describe than uncertainty in the age of a fossil, plus the waiting time between the fossil and the speciation that generated it. Critically, it is important to account for fossil age uncertainty in FBD studies, as not doing so can lead to inaccurate inference of tree topologies, divergences times, or both (Barido-Sottani et al., 2019, 2020b).

However, the fossilized birth–death model still contains parameters for which it may be difficult to choose reasonable values. It is generally known that a small proportion of life that has ever existed has fossilized. But what should the fossilization rate in any particular clade be? Should it change over time? Model selection has long been considered an important part of phylogenetic inference (Zwickl and Holder, 2004; Allman and Rhodes, 2008; Baele and Lemey, 2013). But in the absence of easy-to-use selection software (such as the seminal software for molecular model testing, modelTest; Posada and Crandall (1998)), this practice has not been as widely used in other areas of systematic research, particularly for divergence time estimation (Duchêne et al., 2015). Here, we have used hierarchical model selection to fit a model for each of the FBD's component submodels. For each subcomponent, we competed plausible models. The winning models were combined into a final analysis. Using stepping-stone model estimation, we were able to calculate precise model likelihoods and use them to compare models using the log Bayes Factor.

While this methodology is computationally intensive, it was also tractable. Because no time tree can be inferred without the model to infer a tree first, we first chose our model of morphological evolution. This is also the least computationally intense part of the estimation, and can be completed in a few hours. Using this model, we then chose a clock model, testing four different models (see the next section for a discussion of these models). Finally, using our morphological evolution and clock models, we competed several versions of the FBD model, including three skyline models. Scoring a precise marginal likelihood for the total tripartite model is the most computationally intensive part of the work. By saving this for last, and first fitting the less parameter-rich morphological evolution and clock models, this estimation is made far more tractable for an average researcher to conduct on a laptop or desktop computer.

5.2 What Does the Chosen Set of Models Tell Us?

Being able to fit a model doesn't mean that fitting that model tells us anything about biology. Ideally, we will use our knowledge of the system to turn that model fit into knowledge. As shown in Figure 3, we competed several different models of morphological evolution, clock rate distribution across the tree, and the tree model. Each of these models and parameters has meaning in terms of evolution. As the biological or geological interpretation of these phylogenetic models may not be intuitive to geologist readers, we will now examine what we have learned about evolution from this exercise.

The model of morphological evolution is intended to capture how the phylogenetic characters have evolved over time. It is the chief source of information about the topology. The models of morphological evolution we used were all based on the Mk model (Lewis, 2001). In this model, it is assumed that characters can be in any one of k known character states, that each character can change instantly along a branch, and that probability of change between any two states is equiprobable. In our analyses, the model featuring rate variation among characters fits the dataset better than a single rate of evolution across a dataset. This is somewhat unsurprising, as most work in this group has been conducted under parsimony, a model that assumes each character has its own rate of evolution. We also investigated partitioning in this dataset. Prior work has been equivocal about whether "ecological" traits such as feeding structures evolve at higher rates than do other characters, with nearly equal numbers of studies contradicting the notion (Foote, 1994; Sánchez-Villagra and Williams, 1998; Ciampaglio, 2002) and supporting it (Wagner, 1995; Blomberg et al., 2003; Hopkins and Smith, 2015). Here, we find support for these characters being partitioned, with their own rate heterogeneity parameters, though neither set of characters showed consistently higher rates of evolution.

We examined four clock models. The first was a strict clock. These types of clocks are rarely supported in molecular systematics. Rates of molecular evolution are impacted by generation times, metabolic rate, and mutation rate. For a more in-depth review of this concept, see Warnock and Wright (2020) in this issue. How this translates to rates of morphological evolution over time is not well-studied, but all the above factors also likely impact the evolution of anatomical form. In general, little correspondence has been discovered between molecular and morphological rates of evolution (Bromham et al., 2002). Therefore, the lack of support for a strict clock in our data is unsurprising.

The remaining three clock models describe more biologically interesting scenarios. An autocorrelated clock model implies that a descendant branch will have a rate of evolution that is related to the rate of evolution of its ancestor. This is an appealing model, as we would expect that life history traits that effect possible rates of change may accumulate variation slowly and be similar to their ancestors. We also examined an early burst model in which the rate of evolution slows over time. This, too, is an interesting biological hypothesis that is testable given our data. However, both models were less well-supported than the uncorrelated lognormal clock, a model in which large changes in rates of evolution can be seen among ancestor–descendent pairs. It should be noted that while more flexible in terms of the rate variation allowed between ancestors and descendants, the uncorrelated lognormal is not necessarily the most complex model parametrically. The strength of support for the most flexible model suggests that perhaps there is a substantial amount of variation that is not being captured by our current generations of clock models. There may be a universe of models awaiting description that could be tapped into to fill this need. It is worth noting that in order to compute a clock model, some tree model must be assumed in order to incorporate age information. Therefore, the clock model fitting results shown in Figure 3 assume a time-homogeneous FBD model. It may be worth exploring refitting clock models once the FBD model has been selected. However, this raises issues of circularity in model-fitting procedures that warrant further study.

The final model subcomponent is the tree model. We were able to easily reject a time-homogeneous FBD model in which one rate of speciation, extinction, and fossil sampling applies across the whole tree. The entirety of the tree is only a 12-million-year span of evolutionary history. Being able to reject one model over a relatively small amount of time implies that variable-rate models might be appropriate for a great many systems. Cinctans appeared in a three-stage slice of the Miaolingian Epoch. One competed model had different sets of FBD parameters for the Wuluian, Drumian and Guzhangian stages. Another was a two-stage model in which time was split down the middle in the Drumian. And a third, the most complex model, in which each stage was split into two intervals, was also examined. It is worth noting that discriminate power between these models was fairly good, and that the most complex model was not simply chosen. The three-stage model fit best, followed by the six-stage model, and finally the two-stage model. This is somewhat comforting: if the most complex model had been chosen for each component model, one would wonder if we were not simply choosing from a candidate set of underparameterized models. The Bayes Factor is a reasonably conservative test and did reject more parameter-rich models in favor of simpler ones.

Together, these models paint a picture of evolution in which trophically important characters evolve according to different mechanisms than non-trophic characters. We find evidence for a world in which at times of notable transitions of the Earth (geological stages), we see change in the fundamental processes of diversification and sampling. And we come to understand that from ancestor to descendant, different life history pressures lead to changes in the rate at which evolutionary change accumulates. These first forays into hierarchical model fitting call attention to significant pieces, such as the clock model, where we may be able to examine sources of heterogeneity and improve our models even further.

5.3 Cinctan Phylogeny: Implications for Systematics and Macroevolution

The origin time of the cinctan-*Ctenocystis* group was 507.52 mya (HPD 505.808–508.11 mya), with the ingroup originating at 505.747 mya (HPD 507.27–504.699 mya). As we note in our discussion of the chronostratigraphic data that we use, each tip (i.e., species) has uncertainty associated with its first appearance: we might know that the first occurrence (or possible first occurrences) are in a particular trilobite zone, but that typically restricts the age to a one- to three-million-year window. This might sound trivial if we are thinking about divergences for modern taxa, but here it represents a significant proportion of expected species lifetimes, and thus a potentially large amount of time to accumulate (or not accumulate) character change. In RevBayes' implementation of the FBD model, tip uncertainty is typically treated as a uniform prior between the first and last appearances on the tip taxon (Barido-Sottani et al., 2020a). The uniform prior says that no age within the bounds is *a priori* more likely than any other. Nonetheless, we do see significant structure in the distributions for each tip (Figure A1). Some tips, such as *Ctenocystis* and *Elliptocinctus vizcainoi*, show strong skew toward the older or younger ages within their uniform tip range. Others, such as *Asturicystis jaekeli*, show less signal, retrieving more or less the input uniform prior. This suggests that FBD analyses may be useful in the future for helping to narrow tip age ranges (Drummond and Stadler, 2016; King and Beck, 2020). In clades where tip uncertainty tends to be quite high, this could be an analytical path to higher precision on tip ages.

The topology of the tree is fully resolved but poorly supported on many nodes (Figure 5). This is unsurprising, as bootstrap support values in prior work have also been low (Smith and Zamora, 2009; Zamora et al., 2013a). The placement of *Protocinctus* is interesting on this phylogeny. Although it is the oldest cinctan, *Protocinctus* also possesses some derived character states if we assume that *Ctenocystis* is the appropriate outgroup (Rahman and Zamora, 2009). Accordingly, prior phylogenetic studies place it as a basal member of the Sucocystidae, but evolving after the Sucocystidae diverged from both the Gyrocystidae and Trochocystitidae (Smith and Zamora, 2009; Zamora et al., 2013b). The placement of *Protocinctus* as sister to the rest of the cinctans is likely not solely due to the age of the fossil. This fossil is younger than its parent's next several ancestor nodes, meaning it could have been plausibly placed in a more derived position nested within the cinctan clade, but was not in the phylogeny inferred by the best fitting model (see Figure A2 for alternative positions of *Protocinctus* in suboptimal models). The split between *Protocinctus* and the rest of the cinctans is also one of the most well-supported nodes on this tree. In trees constructed with the second and third best-fit model, *Protocinctus* is sister to a paraphyloetic grouping of Gyrocystidae and Sucocystidae, and a sampled ancestor sister to the Sucocystidae (Figure A2).

In the in-group topology, *Asturicystis*, *Progyrocystis*, and *Graciacystis* form a weakly supported clade that is sister to the rest of the *Gyrocystis*. *Trochocystoides* and *Trochocystites* do not form a monophyletic grouping. Our phylogeny also reflects a closer relationship between *Undatacinctus* and *Ludwigicintus*. Neither *Elliptocinctus* nor *Sucocystis* are monophyletic in this analysis. Some of these differences may represent differences between the model applied here and in previous work. We used the Mk model (Lewis, 2001), which is more robust to superimposed or homoplasious changes than parsimony (Felsenstein, 1978; Wright and Hillis, 2014).

But differences may also reflect the inclusion of age information. For example, *Elliptocinctus* is a genus with two species on this tree, and prior analyses have recovered these as sister taxa. We did not recover *Elliptocinctus* as monophyletic, with *Elliptocinctus barrandei* descending from a node that is 501.865 million years old (age HPD 500.326–503.626 Ma) (Figure 6). This node is younger than the earliest appearance of *Elliptocinctus vizcainoi*. In order for these two taxa to be monophyletic, the strength of evidence in the character data would have to be strong enough to either move *Elliptocinctus vizcainoi* into that grouping (which is a canonical position for *Elliptocinctus*), thereby moving the age of the whole group back several million years, or we would have to move *Elliptocinctus barrandei* out of it. Cinctans have a relatively small amount of characters, and our analyses suggest notable homoplasy in the group. For this reason, the inclusion of other information may be a significant benefit to the accuracy and clarity of phylogenetic and macroevolutionary solutions. In particular, fossil age information is not confounded by homoplasy. Fossil age information is treated as data under the FBD model, which has historically not been true of calibration methods, in which fossil age information was used to set constraints on clade ages. This means that the age information does not directly constrain the topology. Topology is determined from the discrete character data. Fossil age information is used to date the tree and determine which of the topologies are most plausible, given the ages available. It will be worth further exploration to find out when we expect age information to exert a stronger influence than character information in determining a dated tree.

Interestingly, *Gyrocystis* has a number of sampled ancestors in the genus. In this genus, there are a total of three sampled ancestors, one pair of which (*G. erecta* and *G. badulesiensis*) likely represent budding cladogenesis (i.e., a case of speciation where the ancestral species persists). Its sister group also has one, which may also represent evidence for budding speciation. We emphasize that these data show evidence for budding speciation despite the fact that we did not explicitly model stasis nor distinguish between "punctuational" changes associated with speciation versus continuous change within lineages (e.g., Eldredge and Gould, 1972; Wagner and Marcot, 2010). However, the differences between character changes associated with speciation versus continuous background change for these data might be small enough to be subsumed by the uncorrelated relaxed clock model we employed, which models rate shifts as occurring between branches but constant across a given branch's duration. Nevertheless, future analyses employing more complex approaches to modeling

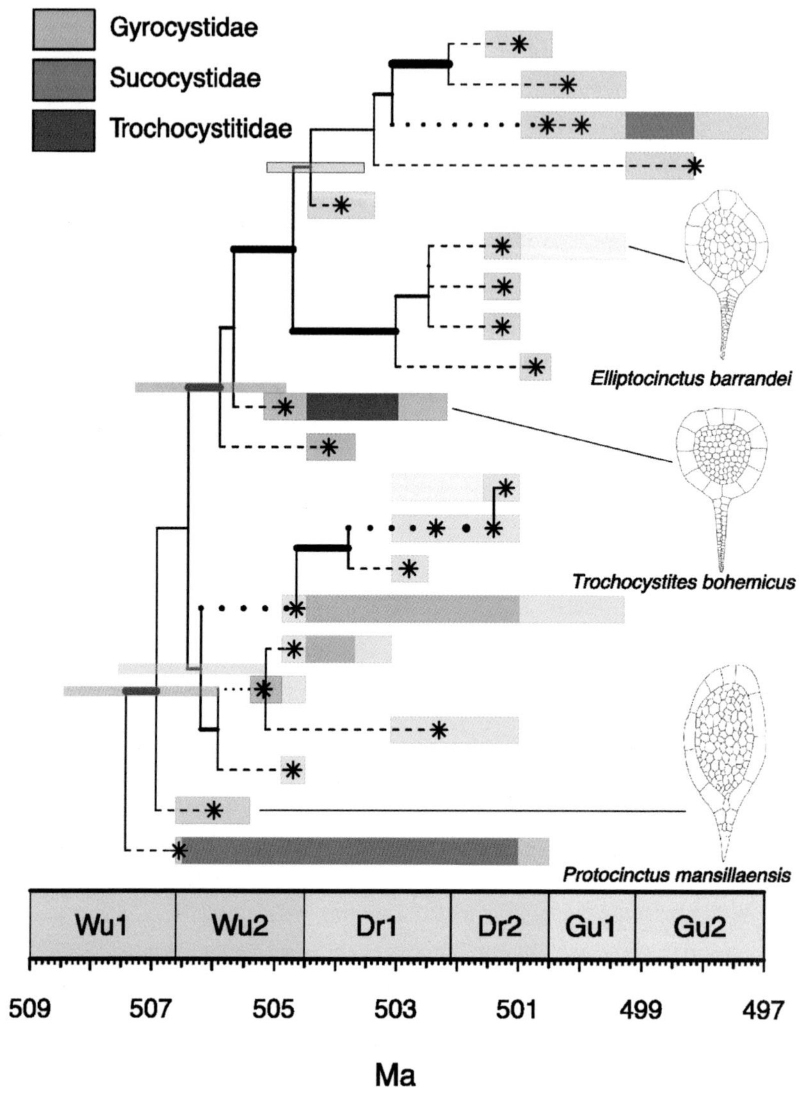

Figure 6 The same phylogeny but with credible intervals given for a few major divergences, includes: (1) cinctans from ctenocystoids (gray); (2) gyrocystids from other cinctans (orange); (3) trochocystitids and sucocystids from other cinctans (purple); (4) sucocystids from trochocystitids. (Because this tree reconstructs all prior familial definitions as polyphyletic, the clades corresponding to these divergences represent plausible groupings for future revisions to cinctan higher taxa.)

speciation dynamics in fossil cinctans may outperform the models considered herein.

Evidence for sampled ancestors in the cinctan fossil record might seem surprising given that the echinoderm record is less complete than that of many other marine invertebrates (Foote and Sepkoski, 1999), although recent FBD studies find strong evidence for their occurrence in the particularly well-sampled record of Paleozoic crinoid echinoderms (Wright, 2017b; Wright and Toom, 2017). However, four of the five cases for cinctans are from the Drumian, for which skyline models imply the highest sampling rate for cinctans. The probability of sampling ancestor–descendant pairs reflects the probability of sampling two species (i.e., [completeness]^2, see Foote, 1996c):

$$Pr[\text{sampling two species}] = \left(\frac{\psi}{\psi + \mu}\right)^2$$

where ψ is the fossilization rate and μ is the extinction rate (Foote, 1997, equation 1b). Still, even at the peak during the Drumian, we expect completeness of 0.21. This in turn predicts that we should find direct ancestor–descendant pairs only 4 percent of the time. However, sampling of ancestors and descendants often should be more probable than global sampling rates imply because sampling rates vary geographically as well as temporally. Because ancestors and descendants usually occur in the same geographic regions and environments, and because ancestors and descendants must at least abut temporally, factors favoring the sampling of any one species often favor the sampling of close relatives, including ancestors (Wagner and Erwin, 1995).

Cinctans have other paleobiological characteristics that make discovery of sampled ancestors relatively probable: (1) the group occurs over a small window of time, allowing for the assessment of taxonomic completeness, (2) they are marine taxa, allowing for better fossilization potential than many groups such as terrestrial vertebrates, (3) they have mineralized skeletons, and are frequently preserved well enough (often as either molds or recrystallized calcite) to code morphological features, (4) they are numerically abundant fossils, particularly in rocks from the Iberian Chains of Spain and the Montagne Noire of southern France, which enables the collection of multiple specimens and assessment of more complete material, and (5) they are small, making it more tractable to score characters from relatively complete specimens.

One final reason why we might not be surprised to find as many cinctan sampled ancestors as we do stems from the fact that reconstructed ancestors cooccur with reconstructed descendants in some cases (Figures 4–5). Budding cladogenesis is consistent with a variety of allopatric speciation models in which biological traits enhancing preservation probabilities (e.g., broad geographic ranges and long durations) also enhance the probability of leaving daughter taxa (Wagner and Erwin, 1995), all else being equal. This becomes particularly relevant because both kinds of occupancy patterns and sampling among contemporaneous species are typically exponential (Liow, 2013; Wagner and Marcot, 2013; Foote, 2016), with common species having individual sampling probabilities much greater than average. Therefore, if high occupancy is linked with the propensity for generating more daughter species, then we expect a fossil record biased in favor of species that had daughter species. This in turn means that we expect ancestor–descendant pairs to be more common than completeness metrics would predict. Our results suggest cinctans conform to this general model, and other taxa with similar preservation rates and sampling intensity may also contain sampled ancestors.

6 Conclusion

In this contribution, we have laid forth a framework for fitting complex, hierarchical phylogenetic models. We also draw attention to the relationship between macroevolutionary models on which many paleobiological studies focus and their corresponding phylogenetic models. We hope our case study of cinctan echinoderms illustrates how the methodological approaches to phylogenetic paleobiology discussed herein provide useful tools for a diverse range of paleontological interests and pursuits. The fossilized birth–death represents a significant leap forward in terms of the integration of fossils in Bayesian phylogenetic analyses. Under this model, fossils are data, not mere clade constraints. However, to leverage this framework involves fitting multiple submodels to a particular dataset. In doing so, we also inferred that a new dated phylogeny for cinctan echinoderms provided some insight as to how and why this phylogeny differs from prior work, and highlighted its systematic and macroevolutionary implications for cinctan paleobiology. We have highlighted several theoretical and empirical concerns, such as how age information impacts topology and how common sampled ancestors are in empirical datasets, which have major implications for discerning models of speciation in the fossil record. It is our hope that in describing how complex model fitting can be carried out in a tractable way, we will empower more taxonomic empiricists to use the FBD approach we utilize herein with their data. We believe the interplay between theoretical phylogenetics and deep taxonomic knowledge of empirical paleontologists is critically important for generating models that not only help us better understand the peculiarities of our favorite taxonomic groups, but also help unravel generalities in the history of life.

Appendix

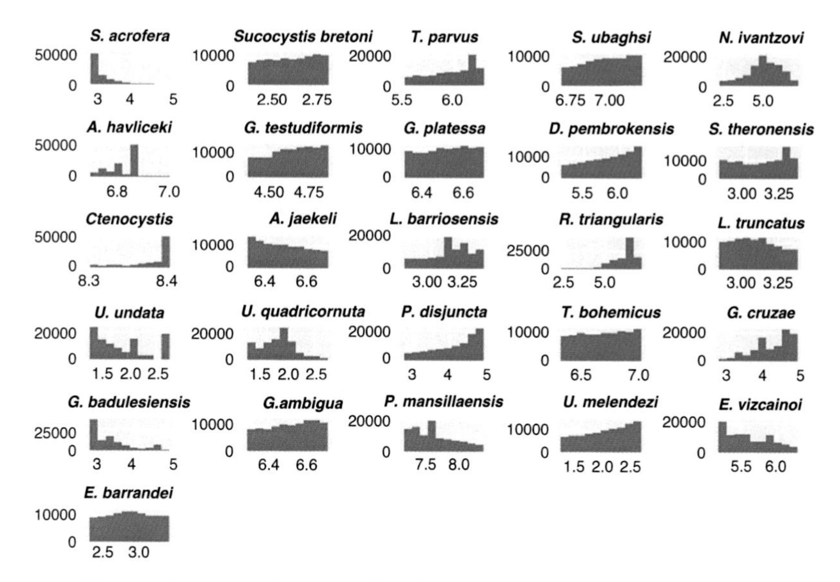

Figure A1 Posterior traces for fossil tip ages. These represent the post-burn-in distributions of ages recovered for each fossil tip. Uncertainty in tip age was parameterized using a normal distribution between the oldest feasible and youngest feasible ages for each fossil tip.

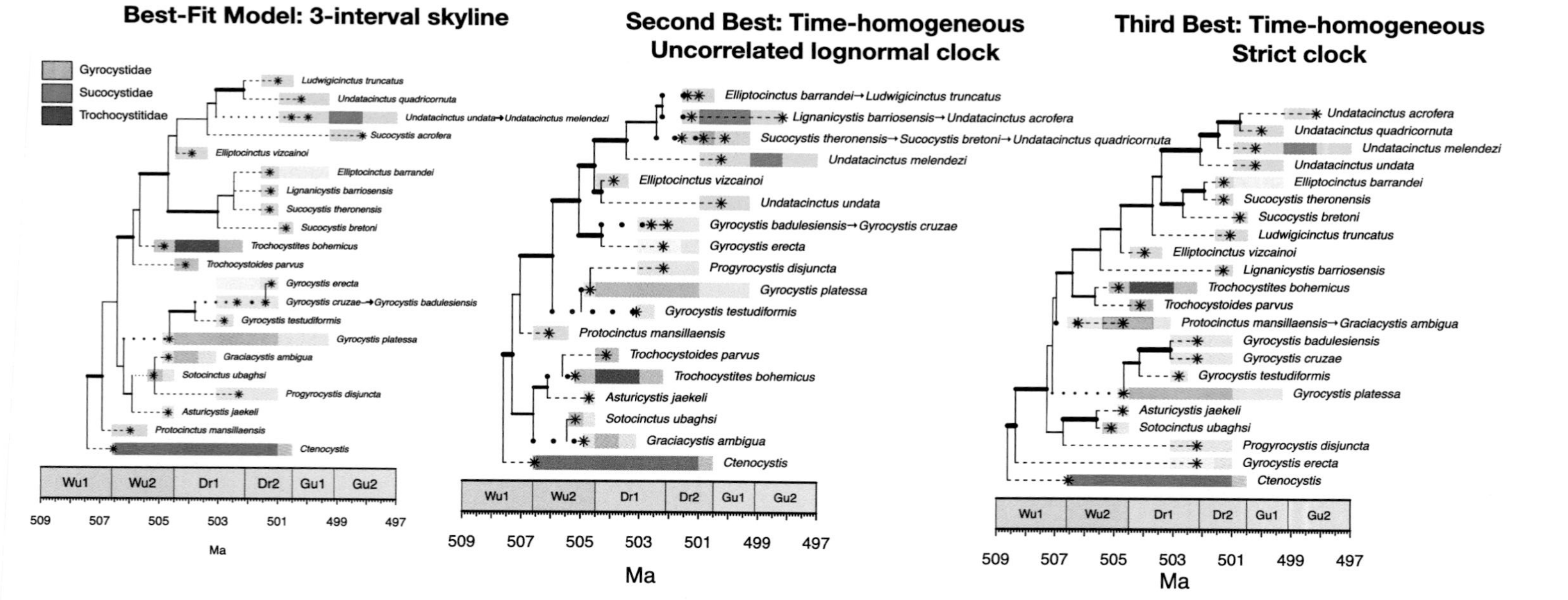

Figure A2 Color-coded phylogenetic trees for the top three best models. As can be seen, families are largely similarly grouped between the five hypotheses. The position of *Protocinctus* varies between the best-fit model and the other two. The number of sampled ancestors also varies among trees. The strict clock tree also contains older divergences internal to the tree.

Table A1 Diversification parameters of the FBD model. Rates presented as the median of the 95 percent HPD of the Bayesian posterior sample for the best-fit model, the model in which each geological stage has its own speciation (λ), extinction (μ), and fossil sampling intensity (ψ) in parameters. Turnover = μ/λ; Diversification=λ-μ. These are the same data as Table 1, but with values in square brackets, indicating the 95 percent HPD of the posterior sample.

Stage	ψ	**Diversification**	λ	μ	**Turnover**
Guzhangian	0.188 [4.044e–3, 0.589]	–0.193 [–35.014, 2.859]	0.494 [6.145E–5, 2.023]	0.687 [4.43E–5, 35.473]	2.148 [.043, 13.044]
Drumian	0.260 [3.023e–3, 1.045]	0.317 [–37.3075, 2.632]	1.28 [2.91E–5, 2.037]	0.964 [1.68E–5, 37.674]	0.714 [7.16E–3, 8.783]
Wuluian	0.224 [3.105e–3, 1.089]	–2.657 [–35.739, 3.181]	0.916 [9.57E–5, 2.016]	3.574 [5.85E–6, 36.524]	4.936 [6.16E–3, 12.493]

References

Allman, E. S., and Rhodes, J. A. 2008. Identifying evolutionary trees and substitution parameters for the general Markov model with invariable sites. *Mathematical Biosciences*, 211:18–33.

Alroy, J. 2015. A more precise speciation and extinction rate estimator. *Paleobiology*, 41(4):633–639. doi: https://doi.org/10.1017/pab.2015.26.

Alvaro, J. J., Lefebvre, B., Shergold, J. H., and Vizcaïno, D. 2001. The Middle–Upper Cambrian of the southern Montagne Noire. *Annales de la Société Géologique du Nord (2e série)*, 8:205–211.

Alvaro, J. J., Ferretti, A., González-Gómez, C., Serpagli, E., Tortello, M. F., Vecoli, M., and Vizcaïno, D. 2007. A review of the Late Cambrian (Furongian) palaeogeography in the western Mediterranean region, NW Gondwana. *Earth-Science Reviews*, 85(1):47–81. doi: https://doi.org/10.1016/j.earscirev.2007.06.006.

Aris-Brosou, S., and Yang, Z. 2002. Effects of models of rate evolution on estimation of divergence dates with special reference to the metazoan 18s ribosomal RNA phylogeny. *Systematic Biology*, 51(5):703–714.

Baele, G., and Lemey, P. 2013. Bayesian evolutionary model testing in the phylogenomics era: matching model complexity with computational efficiency. *Bioinformatics*, 29(16):1970–1979.

Bapst, D. W. 2013. A stochastic rate-calibrated method for time-scaling phylogenies of fossil taxa. *Methods in Ecology and Evolution*, 4(8):724–733. ISSN 2041-210X. doi: https://doi.org/10.1111/2041-210X.12081.

Barido-Sottani, J., Aguirre-Fernández, G., Hopkins, M. J., Stadler, T., and Warnock, R. 2019. Ignoring stratigraphic age uncertainty leads to erroneous estimates of species divergence times under the fossilized birth–death process. *Proceedings of the Royal Society B*, 286(1902):20190685.

Barido-Sottani, J., Justison, J. A., Wright, A. M., Warnock, R. C. M., Pett, W., and Heath, T. A. 2020a. Estimating a time-calibrated phylogeny of fossil and extant taxa using RevBayes. In Céline Scornavacca, Frédéric Delsuc, and Nicolas Galtier, editors, *Phylogenetics in the Genomic Era*, pages 5.2:1–5.2:23. No commercial publisher — Authors' open access book. `https://hal.archives-ouvertes.fr/hal-02536394`.

Barido-Sottani, J., van Tiel, N., Hopkins, M. J., Wright, D. F., Stadler, T., and Warnock, R. C. M. 2020b. Ignoring fossil age uncertainty leads to inaccurate topology and divergence time estimates using time calibrated tree inference.

Frontiers in Ecology and Evolution, 8:123. doi: https://doi.org/10.3389/fevo.2020.00183.

Bergström, S. M., Chen, X., Gutiérrez-Marco, J. C., and Dronov, A. 2009. The new chronostratigraphic classification of the ordovician system and its relations to major regional series and stages and to $\delta^{13}C$ chemostratigraphy. *Lethaia*, 42(1):97–107. doi: https://doi.org/10.1111/j.1502-3931.2008.00136.x.

Blomberg, S. P., Garland, T., Jr, and Ives, A. R. 2003. Testing for phylogenetic signal in comparative data: behavioral traits are more labile. *Evolution*, 57 (4):717–745. doi: https://doi.org/10.1111/j.0014-3820.2003.tb00285.x.

Bottjer, D. J., and Jablonski, D. 1988. Paleoenvironmental patterns in the evolution of post-Paleozoic benthic marine invertebrates. *Palaios*, 3:540–560. doi: https://doi.org/10.2307/3514444.

Bottjer, D. J., Davidson, E. H., Peterson, K. J., and Cameron, R. A. 2006. Paleogenomics of echinoderms. *Science*, 314(5801):956–960.

Bromham, L., Rambaut, A., and Harvey, P. H. 1996. Determinants of rate variation in mammalian DNA sequence evolution. *Journal of Molecular Evolution*, 43(6):610–621.

Bromham, L., Woolfit, M., Lee, M. S. Y., and Rambaut, A. 2002. Testing the relationship between morphological and molecular rates of change along phylogenies. *Evolution*, 56(10):1921–1930. doi: https://doi.org/10.1111/j.0014-3820.2002.tb00118.x.

Bromham, L., Hua, X., Lanfear, R., and Cowman, P. F. 2015. Exploring the relationships between mutation rates, life history, genome size, environment, and species richness in flowering plants. *The American Naturalist*, 185(4): 507–524.

Brusca, R. C., and Brusca, G. J. 2003. *Invertebrates*. Number QL 362. B78 2003. Basingstoke. Sinauer Associates, Sunderland, Massachusetts.

Chlupac, I., Havlicek, V., Kríž, J., Kukal, Z., and Storch, P. 1998. *Palaeozoic of the Barrandian (Cambrian to Devonian)*. Czech Geological Survey, Prague.

Ciampaglio, C. N. 2002. Determining the role that ecological and developmental constraints play in controlling disparity: examples from the crinoid and blastozoan fossil record. *Evolution and Development*, 4(3):170–188. doi: https://doi.org/10.1046/j.1525-142X.2002.02001.x.

Cramer, B. D., Brett, C. E., Melchin, M. J., Männik, P., Kleffner, M. A., McLaughlin, P. I., Loydell, D. K., Munneeke, A., Jeppsson, L., Corradini, C., Brunton, F. R., and Saltzman, M. R. 2011. Revised correlation of Silurian provincial series of North America with global and regional chronostratigraphic units and $\delta^{13}C_{carb}$ chemostratigraphy. *Lethaia*, 44(2):185–202. doi: https://doi.org/10.1111/j.1502-3931.2010.00234.x.

David, B., Lefebvre, B., Mooi, R., and Parsley, R. 2000. Are homalozoans echinoderms? An answer from the extraxial-axial theory. *Paleobiology*, 26 (4):529–555.

Drummond, A. J., and Rambaut, A. 2007. BEAST: Bayesian evolutionary analysis sampling trees. *BMC Evolutionary Biology*, 7:214.

Drummond, A. J., Ho, S. Y. W., Phillips, M. J., and Rambaut, A. 2006. Relaxed phylogenetics and dating with confidence. *PLoS Biology*, 4(5):e88.

Drummond, A. J., and Stadler, T. 2016. Bayesian phylogenetic estimation of fossil ages. *Philosophical Transactions of the Royal Society B: Biological Sciences*, 371(1699):20150129.

Duchêne, D. A., Duchêne, S., Holmes, E. C., and Ho, S. Y. W. 2015. Evaluating the adequacy of molecular clock models using posterior predictive simulations. *Molecular Biology and Evolution*, 32(11):2986–2995.

Eldredge, N. 1971. The allopatric model and phylogeny in paleozoic invertebrates. *Evolution*, 25(1):156–167. doi: https://doi.org/10.2307/2406508.

Eldredge, N., and Gould, S. J. 1972. *Punctuated equilibria: an alternative to phyletic gradualism*, book section 82, pages 82–115. Freeman, San Francisco.

Felsenstein, J. 1978. The number of evolutionary trees. *Systematic Zoology*, 27 (1):27–33. ISSN 00397989.

Felsenstein, J. 1981. Evolutionary trees from DNA sequences: a maximum likelihood approach. *Journal of Molecular Evolution*, 17(6): 368–376.

Foote, M. 1992. Paleozoic record of morphological diversity in blastozoan echinoderms. *Proceedings of the National Academy of Sciences, USA*, 89 (16):7325–7329. doi: https://doi.org/10.1073/pnas.89.16.7325.

Foote, M. 1993. Discordance and concordance between morphological and taxonomic diversity. *Paleobiology*, 19(2):185–204. doi: https://doi.org/10.2307/2400876.

Foote, M. 1994. Morphological disparity in Ordovician–Devonian crinoids and the early saturation of morphological space. *Paleobiology*, 20(3):320–344. doi: https://doi.org/10.2307/2401006.

Foote, M. 1996a. Ecological controls on the evolutionary recovery of post–paleozoic crinoids. *Science*, 274(5292):1492–1495. doi: https://doi.org/10.1126/science.274.5292.1492.

Foote, M. 1996b. *Models of morphologic diversification*, book section 62, pages 62–86. University of Chicago Press, Chicago.

Foote, M. 1996c. On the probability of ancestors in the fossil record. *Paleobiology*, 22(2):141–151. doi: https://doi.org/10.1666/0094-8373-22.2.141.

Foote, M. 2016. On the measurement of occupancy in ecology and paleontology. *Paleobiology*, 42:707–729. doi: https://doi.org/10.1017/pab.2016.24.

Foote, M., and Sepkoski, J. J., Jr. 1999. Absolute measures of the completeness of the fossil record. *Nature*, 398:415–417. doi: https://doi.org/10.1038/18872.

Foote, M. 1997. Estimating taxonomic durations and preservation probability. *Paleobiology*, 23(3):278–300.

Friedrich, W. P. 1993. Systematik und Funktionsmorphologie mittelkambrischer Cincta (Carpoidea, Echinodermata). *Beringeria*, 7.

Gaut, B. S., Muse, S. V., Clark, W. D., and Clegg, M. T. 1992. Relative rates of nucleotide substitution at the rbcl locus of monocotyledonous plants. *Journal of Molecular Evolution*, 35(4):292–303.

Gavryushkina, A., Heath, T. A., Ksepka, D. T., Stadler, T., Welch, D., and Drummond, A. J. 2017. Bayesian total-evidence dating reveals the recent crown radiation of penguins. *Systematic Biology*, 66(1):57–73.

Geyer, G. 2019. A comprehensive Cambrian correlation chart. *International Union of Geological Sciences*, 42(4):321–332. doi: https://doi.org/10.18814/epiiugs/2019/019026.

Geyer, G., and Landing, E. 2006. Ediacaran–Cambrian depositional environments and stratigraphy of the western atlas regions. *Beringeria Special Issue*, 6:47–120.

Geyer, G., and Shergold, J. 2000. The quest for internationally recognized divisions of Cambrian time. *Episodes*, 23(3):188–195.

Gradstein, F. M., Ogg, J. G., and Schmitz, M. 2012. *The geologic time scale 2012*, Volume 2. Elsevier, Amsterdam.

Gradstein, F. M., Ogg, J. G., Schmitz, M. D., and Ogg, G. M. 2020. *Geologic time scale 2020*. Elsevier, Amsterdam. ISBN 978-0-12-824360-2. doi: https://doi.org/10.1016/C2020-1-02369-3.

Harvey, P. H., and Pagel, M. D. 1991. *The comparative method in evolutionary biology*, Volume 239. Oxford University Press, Oxford.

Hasegawa, M., Kishino, H., and Yano, T. 1985. Dating of the human–ape splitting by a molecular clock of mitochondrial DNA. *Journal of Molecular Evolution*, 22(2):160–174.

Heath, T. A., Huelsenbeck, J. P., and Stadler, T. 2014. The fossilized birth–death process for coherent calibration of divergence-time estimates. *Proceedings of the National Academy of Sciences*, 111(29):E2957–E2966.

Hopkins, M. J., and Smith, A. B. 2015. Dynamic evolutionary change in post-Paleozoic echinoids and the importance of scale when interpreting changes in rates of evolution. *Proceedings of the National Academy of Sciences*, 112 (2):3758–3763. doi: https://doi.org/10.1073/pnas.1418153112.

Huelsenbeck, J. P., Larget, B., and Swofford, D. L. 2000. A compound Poisson process for relaxing the molecular clock. *Genetics*, 154:1879–1892.

Hughes, M., Gerber, S., and Wills, M. A. 2013. Clades reach highest morphological disparity early in their evolution. *Proceedings of the National Academy of Sciences*, 110(34):13875–13879. doi: https://doi.org/10.1073/pnas.1302642110.

Jablonski, D. 2020. *Macroevolutionary theory*, book section 338, pages 338–368. University of Chicago Press, Chicago.

Kass, R. E., and Raftery, A. E. 1995. Bayes factors. *Journal of the American Statistical Association*, 90:773–795.

Kimura, M. 1980. A simple method for estimating evolutionary rates of base substitutions through comparative studies of nucleotide sequences. *Journal of Molecular Evolution*, 16(2):111–120.

King, B., and Beck, Robin M. D. 2020. Tip dating supports novel resolutions of controversial relationships among early mammals. *Proceedings of the Royal Society B: Biological Sciences*, 287(1928):20200943. doi: https://doi.org/10.1098/rspb.2020.0943.

LaPolla, J. S., Dlussky, G. M., and Perrichot, V. 2013. Ants and the fossil record. *Annual Review of Entomology*, 58(1):609–630. doi: https://doi.org/10.1146/annurev-ento-120710-100600.

Lefebvre, B., Guensburg, T. E., Martin, E. L., Rich, M., Elise, N., Nohejlová, M., Saleh, F., Kouraïss, K., Khadija, E. H., and David, B. 2019. Exceptionally preserved soft parts in fossils from the Lower Ordovician of Morocco clarify stylophoran affinities within basal deuterostomes. *Geobios*, 52:27–36.

Lewis, P. O. 2001. A likelihood approach to estimating phylogeny from discrete morphological character data. *Systematic Biology*, 50(6):913–925.

Liñán, E., Perejón, A., Gozalo, R., Moreno-Eiris, E., and de Oliveira, J. T. 2004. The Cambrian system in Iberia. *Cuadernos del Museo Geominero*, 3: 1–63.

Liow, L. H. 2013. Simultaneous estimation of occupancy and detection probabilities: an illustration using Cincinnatian brachiopods. *Paleobiology*, 39(2): 193–213. ISSN 0094-8373. doi: https://doi.org/10.1666/12009.

Liow, L. H., Quental, T. B., and Marshall, C. R. 2010. When can decreasing diversification rates be detected with molecular phylogenies and the fossil record? *Systematic Biology*, 59(6):646.

Marshall, C. R. 2017. Five palaeobiological laws needed to understand the evolution of the living biota. *Nature Ecology Evolution*, 1(6):0165. doi: https://doi.org/10.1038/s41559-017-0165.

Nardin, E., Lefebvre, B., Fatka, O., Nohejlová, M., Kašička, L., Šinágl, M., and Szabad, M. 2017. Evolutionary implications of a new transitional blastozoan echinoderm from the Middle Cambrian of the Czech Republic. *Journal of Paleontology*, 91(4):672–684. doi: https://doi.org/10.1017/jpa.2016.157.

Nichols, D. 1972. The water-vascular system in living and fossil echinoderms. *Palaeontology*, 15:519–538.

Nowak, M. D., Smith, A. B., Simpson, C., and Zwickl, D. J. 2013. A simple method for estimating informative node age priors for the fossil calibration of molecular divergence time analyses. *PLoS One*, 8(6):e66245.

Nylander, J. A. A., Ronquist, F., Huelsenbeck, J. P., and Nieves-Aldrey, J. 2004. Bayesian phylogenetic analysis of combined data. *Systematic Biology*, 53(1): 47–67.

Poinar, G. O., and Mastalerz, M. 2000. Taphonomy of fossilized resins: determining the biostratinomy of amber. *Acta Geologica Hispanica*, 35(1): 171–182.

Posada, D., and Crandall, K. A. 1998. Modeltest: testing the model of dna substitution. *Bioinformatics (Oxford, England)*, 14(9): 817–818.

Quental, T. B., and Marshall, C. R. 2009. Extinction during evolutionary radiations: reconciling the fossil record with molecular phylogenies. *Evolution*, 63(12):3158–3167.

Quental, T. B., and Marshall, C. R. 2010. Diversity dynamics: molecular phylogenies need the fossil record. *Trends in Ecology & Evolution*, 25: 434–441.

Rahman, I. A., Zamora, S., Falkingham, P. L., and Phillips, J. C. 2015. Cambrian cinctan echinoderms shed light on feeding in the ancestral deuterostome. *Proceedings of the Royal Society B: Biological Sciences*, 282(1818): 20151964.

Rahman, I. A. 2009. Making sense of carpoids. *Geology Today*, 25(1):34–38.

Rahman, I. A. 2016. Fossil focus: Cinctans. *Palaeontology Online*, 6(4):1–7.

Rahman, I. A., O'Shea, J., Lautenschlager, S., and Zamora, S. 2020. Potential evolutionary trade-off between feeding and stability in Cambrian cinctan echinoderms. *Palaeontology*, 63(5):689–701. ISSN 0031-0239. doi: https://doi.org/10.1111/pala.12495.

Rahman, I. A., Sutton, M. D., and Bell, M. A. 2009. Evaluating phylogenetic hypotheses of carpoids using stratigraphic congruence indices. *Lethaia*, 42 (4):424–437. doi: https://doi.org/10.1111/j.1502-3931.2009.00161.x.

Rahman, I. A., and Zamora, S. 2009. The oldest cinctan carpoid (stem-group echinodermata), and the evolution of the water vascular system. *Zoological Journal of the Linnean Society*, 157(2):420–432. doi: https://doi.org/10.1111/j.1096-3642.2008.00517.x.

Rambaut, A., Drummond, A. J., Xie, D., Baele, G., and Suchard, M. A. 2018. Posterior summarization in Bayesian phylogenetics using Tracer 1.7. *Systematic Biology*, 67(5):901–904. ISSN 1063-5157. doi: https://doi.org/10.1093/sysbio/syy032.

Rasmussen, C. M. Ø., Kröger, B., Nielsen, M. L., and Colmenar, J. 2019. Cascading trend of Early Paleozoic marine radiations paused by Late Ordovician extinctions. *Proceedings of the National Academy of Sciences*, 116 (15):7207–7213. doi: https://doi.org/10.1073/pnas.1821123116.

Sánchez-Villagra, M. R., and Williams, B. A. 1998. Levels of homoplasy in the evolution of the mammalian skeleton. *Journal of Mammalian Evolution*, 5 (2):113–126. doi: https://doi.org/10.1023/A:1020549505177.

Sanderson, M. J. 2002. Estimating absolute rates of molecular evolution and divergence times: a penalized likelihood approach. *Molecular Biology and Evolution*, 19(1):101–109.

Sdzuy, K. 1993. Early cincta (carpoidea) from the Middle Cambrian of Spain. *Beringia*, 8:189–207.

Sepkoski, J. J., Jr. 1981. A factor analytic description of the Phanerozoic marine fossil record. *Paleobiology*, 7(1):36–53. doi: https://doi.org/10.1017/S0094837300003778.

Sheffield, S. L., and Sumrall, C. D. 2019. The phylogeny of the diploporita: a polyphyletic assemblage of blastozoan echinoderms. *Journal of Paleontology*, 93(4):740–752.

Smith, A. B., and Zamora, S. 2009. Rooting phylogenies of problematic fossil taxa: a case study using cinctans (stem-group echinoderms). *Palaeontology*, 52(4):803–821. doi: https://doi.org/10.1111/j.1475-4983.2009.00880.x.

Smith, A. B., and Zamora, S. 2013. Cambrian spiral-plated echinoderms from Gondwana reveal the earliest pentaradial body plan. *Proceedings of the Royal Society B: Biological Sciences*, 270(1765):20131197.

Smith, A. B., Zamora, S., and Álvaro, J. J. 2013. The oldest echinoderm faunas from Gondwana show that echinoderm body plan diversification was rapid. *Nature Communications*, 4(1):1–7.

Smith, A. B. 2005. The pre-radial history of echinoderms. *Geological Journal*, 40(3):255–280.

Smith, A. B., and Swalla, B. J. 2009. Deciphering deuterostome phylogeny: molecular, morphological, and palaeontological perspectives. In

M. J. Telford and D. T. J. Littlewood, editors, *Animal Evolution: Genomes, Fossils and Trees*, pages 80–92. Oxford University Press, Oxford.

Sprinkle, J., and Collins, D. 2006. New eocrinoids from the Burgess Shale, southern British Columbia, Canada, and the Spence Shale, northern Utah, USA. *Canadian Journal of Earth Sciences*, 43(3):303–322. doi: https://doi.org/10.1139/e05-107.

Sprinkle, J. 1973. *Morphology and evolution of blastozoan echinoderms*. Harvard University, Museum of Comparative Zoology, Special Publication, Cambridge, MA, pages 1–283.

Sprinkle, J., and Kier, P. M. 1987. *Phylum Echinodermata*, pages 550–611. In Boardman, R. S., Cheetham, A. H., and Rowell, A. J. (eds.), *Fossil Invertebrates*. Blackwell Scientific Publications, Palo Alto, California

Stadler, T. 2011. Mammalian phylogeny reveals recent diversification rate shifts. *Proceedings of the National Academy of Sciences*, 108(15): 6187–6192.

Stadler, T., Kühnert, D., Bonhoeffer, S., and Drummond, A. J. 2013. Birth–death skyline plot reveals temporal changes of epidemic spread in HIV and hepatitis C virus (HCV). *Proceedings of the National Academy of Sciences*, 110(1):228–233.

Sumrall, C. D. 1997. The role of fossils in the phylogenetic reconstruction of echinodermata. *The Paleontological Society Papers*, 3:267–288.

Sumrall, C. D., and Waters, J. A. 2012. Universal elemental homology in glyptocystitoids, hemicosmitoids, coronoids and blastoids: steps toward echinoderm phylogenetic reconstruction in derived blastozoa. *Journal of Paleontology*, 86:956–927.

Tavaré, S. 1986. Some probabilistic and statistical problems in the analysis of DNA sequences. *Some Mathematical Questions in Biology: DNA Sequence Analysis*, 17:57–86.

Termier, H., and Termier, G. 1973. Les echinodermes cincta du cambriende la Montagne Noire (France). *Geobios*, 6(4):243–265. doi: https://doi.org/10.1016/S0016-6995(73)80019-1.

Thomas, J. A., Welch, J. J., Woolfit, M., and Bromham, L. 2006. There is no universal molecular clock for invertebrates, but rate variation does not scale with body size. *Proceedings of the National Academy of Sciences*, 103(19): 7366–7371.

Valentine, J. W. 1980. Determinants of diversity in higher taxonomic categories. *Paleobiology*, 6(4):444–450.

Valentine, J. W. et al. 1969. Patterns of taxonomic and ecological structure of the shelf benthos during Phanerozoic time. *Palaeontology*, 12(4):684–709.

Wagner, P. 2021. PaleoDB for RevBayes Webinar: PBDB for RevBayes v01.0. January. doi: https://doi.org/10.5281/zenodo.4426555.

Wagner, P. J. 1995. Testing evolutionary constraint hypotheses with early paleozoic gastropods. *Paleobiology*, 21(3):248–272. doi: https://doi.org/10.2307/2401166.

Wagner, P. J. 2019. On the probabilities of branch durations and stratigraphic gaps in phylogenies of fossil taxa when rates of diversification and sampling vary over time. *Paleobiology*, 28(1):30–55. doi: https://doi.org/10.1017/pab.2018.35.

Wagner, P. J., and Erwin, D. H. 1995. *Phylogenetic patterns as tests of speciation models*, book section 87, pages 87–122. Columbia University Press, New York.

Wagner, P. J., and Estabrook, G. F. 2015. The implications of stratigraphic compatibility for character integration among fossil taxa. *Systematic Biology*, 64(5):838–852. doi: https://doi.org/10.1093/sysbio/syv040.

Wagner, P. J., and Marcot, J. D. 2010. *Probabilistic phylogenetic inference in the fossil record: current and future applications*, Volume 16, book section 195, pages 195–217. Paleontological Society, New Haven, CT.

Wagner, P. J., and Marcot, J. D. 2013. Modelling distributions of fossil sampling rates over time, space and taxa: assessment and implications for macroevolutionary studies. *Methods in Ecology and Evolution*, 4(8):703–713. doi: https://doi.org/10.1111/2041-210X.12088.

Warnock, R. C., and Wright, A. M. 2021. *Understanding the tripartite approach to Bayesian divergence time estimation.* Cambridge University Press.

Wright, A. M., and Hillis, D. M. 2014. Bayesian analysis using a simple likelihood model outperforms parsimony for estimation of phylogeny from discrete morphological data. *PLoS One*, 9(10):e109210.

Wright, A. M., Lloyd, G. T., and Hillis, D. M. 2016. Modeling character change heterogeneity in phylogenetic analyses of morphology through the use of priors. *Systematic Biology*, 65(4):602–611.

Wright, D. F. 2017a. Phenotypic innovation and adaptive constraints in the evolutionary radiation of Palaeozoic crinoids. *Scientific Reports*, 7(1):13745. doi: https://doi.org/10.1038/s41598-017-13979-9.

Wright, D. F. 2017b. Bayesian estimation of fossil phylogenies and the evolution of Early to Middle Paleozoic crinoids (echinodermata). *Journal of Paleontology*, 91(4):799–814.

Wright, D. F., Ausich, W. I., Cole, S. R., Peter, M. E., and Rhenberg, E. C. 2017. Phylogenetic taxonomy and classification of the crinoidea (echinodermata). *Journal of Paleontology*, 91(4):829–846.

Wright, D. F., and Toom, U. 2017. New crinoids from the Baltic region (Estonia): fossil tip-dating phylogenetics constrains the origin and Ordovician–Silurian diversification of the flexibilia (echinodermata). *Palaeontology*, 60(6):893–910.

Xie, W., Lewis, P. O., Fan, Y., Kuo, L., and Chen, M. H. 2011. Improving marginal likelihood estimation for Bayesian phylogenetic model selection. *Systematic Biology*, 60(2):150–160.

Zamora, S. 2009. *Equinodermos del Cámbrico medio de las Cadenas Ibéricas y de la zona Cantábrica (Norte de España)*. Thesis.

Zamora, S., and Álvaro, J. J. 2010. Testing for a decline in diversity prior to extinction: Languedocian (latest mid-Cambrian) distribution of cinctans (echinodermata) in the Iberian Chains, NE Spain. *Palaeontology*, 56(6): 1349–1368.

Zamora, S., Lefebvre, B., Álvaro, J. J., Clausen, S., Elicki, O., Fatka, O., Jell, P., Kouchinsky, A., Lin, J. P., Nardin, E., and Parsley, R. 2013a. Cambrian echinoderm diversity and palaeobiogeography. *Geological Society of London, Memoirs*, 38(1):157–171.

Zamora, S., Rahman, I. A., and Smith, A. B. 2012. Plated Cambrian bilaterians reveal the earliest stages of echinoderm evolution. *PLoS One*, 7(6):e38296.

Zamora, S., and Smith, A. B. 2008. A new Middle Cambrian stem-group echinoderm from Spain: Palaeobiological implications of a highly asymmetric cinctan. *Acta Palaeontologica Polonica*, 53(2):207–220.

Zamora, S., and Rahman, I. A. 2014. Deciphering the early evolution of echinoderms with Cambrian fossils. *Palaeontology*, 57(6): 1105–1119.

Zamora, S., and Rahman, I. A. 2015. Palaeobiological implications of a mass-mortality assemblage of cinctans (echinodermata) from the Cambrian of Spain. In S. Zamora and I. Rábano, editors, *Progress in Echinoderm Paleobiology*, pages 203–206. Instituto Geológico y Minero de España.

Zamora, S., Rahman, I. A., and Smith, A. B. 2013b. The ontogeny of cinctans (stem-group echinodermata) as revealed by a new genus, graciacystis, from the Middle Cambrian of Spain. *Palaeontology*, 56(2):399–410. doi: https://doi.org/10.1111/j.1475-4983.2012.01207.x.

Zamora, S., Wright, D. F., Mooi, R., Lefebvre, B., Guensburg, T. E., Gorzelak, P., David, B., Sumrall, C. D., Cole, S. R., Hunter, A. W., Sprinkle, J., Thompson, J. R., Ewin, T. A. M., Fatka, O., Nardin, E., Reich, M., Nohejlová, M., and Rahman, I. 2020. Re-evaluating the phylogenetic position of the enigmatic Early Cambrian deuterostome yanjiahella. *Nature Communications*, 11(1):1286.

Zwickl, D. J., and Holder, M. T. 2004. Model parameterization, prior distributions, and the general time-reversible model in Bayesian phylogenetics. *Systematic Biology*, 53(6):877–888.

Acknowledgments

Supporting scripts can be found in the supplemental code package DOI:10.5281/zenodo.4421405. This Element originated as one of several manuscripts associated with the Paleontological Society short course hosted at the Geological Society Annual Meetings in Phoenix. We thank the Paleontological Society for the support that made the course and manuscript possible. We thank A. Lin, P. Novack-Gottshall, P. Borkow, P. Hearn, and A. Hendy for contributing substantial amounts of the Paleobiology Database information that we use, and S. R. Cole for assistance with obtaining line drawings for fossil cinctans. S. Carlson, S. Zamora, B. Deline, and one anonymous reviewer are thanked for their insightful and constructive reviews. A. M. W. acknowledges support from an Institutional Development Award (IDeA) from the National Institute of General Medical Sciences of the National Institutes of Health under grant number P2O GM103424-18. D. F. W. acknowledges support from the Gerstner Scholars Fellowship and the Gerstner Family Foundation, the Lerner-Gray Fund for Marine Research, and the Richard Gilder Graduate School, American Museum of Natural History, as well as a Norman Newell Early Career Grant from the Paleontological Society. This is PBDB Publication 397.

Cambridge Elements

Elements of Paleontology

About the Series

The Elements of Paleontology series is a publishing collaboration between the Paleontological Society and Cambridge University Press. The series covers the full spectrum of topics in paleontology and paleobiology, and related topics in the Earth and life sciences of interest to students and researchers of paleontology.

The Paleontological Society is an international nonprofit organization devoted exclusively to the science of paleontology: invertebrate and vertebrate paleontology, micropaleontology, and paleobotany. The Society's mission is to advance the study of the fossil record through scientific research, education, and advocacy. Its vision is to be a leading global advocate for understanding life's history and evolution. The Society has several membership categories, including regular, amateur/avocational, student, and retired. Members, representing some 40 countries, include professional paleontologists, academicians, science editors, Earth science teachers, museum specialists, undergraduate and graduate students, postdoctoral scholars, and amateur/avocational paleontologists.

Cambridge Elements $^{\equiv}$

Elements of Paleontology

Elements in the Series

A full series listing is available at: www.cambridge.org/EPLY

Printed in the United States
by Baker & Taylor Publisher Services